Linear Algebra

Lorenzo Robbiano

Linear Algebra

for everyone

 Springer

Lorenzo Robbiano
Dipartimento di Matematica
Università di Genova, Italia

Translated by **Anthony Geramita**
Department of Mathematics and Statistics, Queen's University, Kingston, Canada
and Dipartimento di Matematica, Università di Genova, Italia
From the original Italian edition:
Lorenzo Robbiano, ALGEBRA LINEARE per tutti. © Springer-Verlag Italia 2007

ISBN 978-88-470-1838-9   ISBN 978-88-470-1839-6 (eBook)
DOI 10.1007/978-88-470-1839-6

Library of Congress Control Number: 2010935460

Springer Milan Dordrecht Heidelberg London New York

© Springer-Verlag Italia 2011

Cover-Design: Simona Colombo, Milano

Typesetting: PTP-Berlin, Protago TeX-Production GmbH, Germany (www.ptp-berlin.eu)
Printing and Binding: Signum srl, Bollate (MI)

Springer-Verlag Italia srl – Via Decembrio 28 – 20137 Milano
Springer is a part of Springer Science+Business Media (www.springer.com)

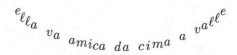

(a palindromic verse[1] written by the author and painted
on a sundial at his home in Castelletto d'Orba

from PALINDROMI DI (LO)RENZO
by Lorenzo)

dedicated to those who read this dedication,
in particular to G

---

[1]lightly she goes from hilltop to valley

Foreword

the moon is more useful that the sun
since at night there is more need of light

(Mullah Nasrudin)

Once (upon a time?) there was a city nestled between the sea and the mountains where the university was divided into faculties. One of these was the Faculty of Science where various subjects were taught; subjects such as Mathematics, Computer Science, Statistics, Physics, Chemistry, Biology, Geology, Environmental Studies and some other things. These subjects were divided into such a myriad of courses that one could easily lose count of them. The only invariant, one of the few things that gave some unity to all of these areas of study, was the fact that every one of these subjects included some mathematics in their introductory courses. This meant that all the students who were enrolled in this faculty would, sooner or later, encounter some of the notions of linear algebra.

Everyone who taught in the Faculty of Science knew that this material was at the base of the scientific pyramid and all were conscious of the fact that no scientist could call himself such if he or she were unable to master the technical fundamentals of linear algebra. However tradition, mixed with convenience, had created courses in each of the various departments of the faculty which contained some basic notions of mathematics but were totally different from each other. As a consequence, it was perfectly possible for a student in Biology to be ignorant of certain fundamental facts of linear algebra which were, however, taught to students of Geology.

Then something unexpected happened. On the day... of the year... there was a meeting of some wise professors from the Faculty of Science (and also from some other Faculties where mathematics was taught). The purpose of this meeting was to correct the situation described above and, after ample and articulate discussion (for that is how it is reported in the minutes of that meeting) it was unanimously decided to assign to a mathematician the task of writing a book of *linear algebra for everyone*.

Some historians claim that the decision was indeed *not* unanimous and that the minutes were altered afterwards. Some assert that the writing of that book was, in fact, proposed by a mathematician and that he was not even supported by his mathematical colleagues. There are even some revisionist historians who assert that the meeting never took place! Perhaps the question needs to be studied further, but one thing is sure: the book was written.

la luna e la terra non sono sole[1]

(from THE BOOK OF SURE THINGS)

---

[1]Some sentences are not translated because the sense would be completely lost.

Introduction

numbers, symbols, algorithms,
theorems,
algorithms, symbols, numbers

(Indrome Pal)

Where does this strange title *Linear Algebra for everyone* come from? In what sense does the author mean "everyone"? It is, of course, common knowledge that mathematics is not for *everyone*. Moreover, it is also often true that one of the obstacles to the dissemination of mathematics are the mathematicians themselves (fortunately not *every one* of them). Indeed, some mathematicians love to play the role of *gate keepers* developing a language which is abstruse and sometimes incomprehensible even to the experts in their own field!

But, suppose we asked a professional mathematician to step back a bit from his habitual way of speaking and write in a more *linear* fashion? And suppose we even asked more, for example, that he make his writing lively? And, since we are asking for so much, suppose we were to ask that the writing even be entertaining? That would not be an easy job, since as a proverb says "few sage things are said lightly while many stupid things are said seriously".

The purpose of this book is to furnish the reader with the first mathematical tools needed to understand one of the pillars of modern mathematics, i.e. *linear algebra*. The text has been written by a mathematician who has tried to step out of his usual character in order to speak to a larger public. He has also taken up the challenge of trying to make accessible *to everyone* the first ideas and the first techniques of a body of knowledge that is fundamental to all of science and technology.

Like a good photographer, the author has tried to create a geometric and chromatic synthesis. Like an able dancer he has tried to merge the solidity of his steps with the lightness of his movements. And, like an expert horticulturist he has tried to always maintain a healthy level of reality. The author has had success in an endeavor such as this. He has, through his leadership of one of the research groups at the University of Genoa, been active in the development of the programme, called CoCoA (see [Co]), which took the very abstract ideas of symbolic calculation and made them easy to use.

But, does the author really believe that *everyone* will read what he has written? In fact, although the book is declared to be *for everyone* it is difficult to imagine, for example, that more than a few retired people or housewives would be able to read it beyond the first few pages. On the other hand, it would not be a bad idea for this book to be read carefully by all university students who have at least one course of mathematics in their programme.

Thus, this book should be read by *at least* students of statistics, engineering, physics, chemistry, biology, natural science, medicine, law... As for mathematics students... Why not? It certainly would not do them any harm to see the fundamentals of linear algebra presented a bit differently, and in a more motivated way, than they would probably find in many, so-called, canonical texts.

Naturally, mathematicians, and in particular algebraists, will observe immediately that the book lacks a formal underpinning. They will notice the fact that definitions, theorems and proofs, in other words all the formal baggage which permeates modern texts of mathematics, are almost totally absent from this book. Perhaps they would enjoy *definitions* like those of Bob Hope, according to whom

a bank is: *the place where they lend you money*
if you can prove you don't need it

or perhaps that of the anonymous author according to whom

modern man is: *the missing link*
between apes and human beings

The author could find his way out of this *a la Hofstadter*, saying that the book contains all the formalism necessary *only when it is closed*. In fact, this author thinks that whatever choices one makes should be made very clear at the outset. In this case, the fundamental choice was that of writing a book *for everyone* and thus with a style and language as close as possible to that which is used everyday.

There is an Indian proverb, which says that an *ounce of practice* is worth more than a ton of theory. Hence, another fundamental decision made by the author was that of proposing hundreds of exercises of varying difficulty to the reader. Some of these, about 2 dozen in all, require the use of a calculator for their solution. Let me clarify what I mean: we know that some readers might be particularly adept at solving problems by hand and get great satisfaction out of doing so with a system of linear equations with lots of equations and unknowns. It's not my intention to deny anyone that pleasure, but, it is best to understand that such an undertaking is essentially useless.

We live, today, in an era in which high speed calculators and excellent programmes are as available to us as pencil and paper. It is best to learn how to use such things well and, at the same time, absolutely fundamental to understand how they work – even at the price of contradicting Picasso.

calculators are useless,
they can only give you answers

(Pablo Picasso)

Let's return, for a moment, to our description of these special exercises. First, they are all marked with the symbol @ so you will understand immediately the kind of problem they are. What can you do to solve them? You will not be left all alone to deal with these problems. In the Appendix, at the end of this volume, you will find explanations and suggestions for solving these types of problems and you will also be shown some explicit solutions using CoCoA (see [Co]). What is CoCoA?

As was already mentioned, CoCoA is a system of symbolic calculation which has been developed by a group of researchers in the Department of Mathematics of the University of Genoa, led by the author. To find out more about this system, the reader is invited to read the Appendix but, better yet, to consult the web page

http://cocoa.dima.unige.it

We now come to the book's organization. The book is divided into a preparatory part (which contains this introduction and the index), an initial chapter with introductory material, two mathematically essential parts (each divided into four chapters) and some concluding remarks. The appendix, of which we have just spoken, is found in this last part as well as some *presumed* conclusions and some bibliographic references. The arrangement of the more mathematical contents is made so that the initial chapter and the first part of the book can serve as a very leisurely introduction to the material. Here the reader is taken by the hand and accompanied gradually through the main themes of linear algebra.

The most important instrument we will use is the systematic discussion of examples. In fact, it is even written in fortune cookies that *the best gift one can bestow on others is a good example.* The central role is played by the objects called matrices, which first enter very unassumingly on the scene and progressively reveal their many facets and their adaptability to situations and problems which are surprisingly diverse. The reader is brought along to understand the significance of a *mathematical model* and of *computational costs.* This is the part that is truly *for everyone* and it was from this that the title of the book was taken.

The second part is still *for everyone*, provided... the reader has understood the first part and no longer needs to be accompanied by hand. As in the first part, the examples still play a major role but the concepts begin to get a bit more elaborate. Coming on stage now are characters which are a bit more complex, at times even *spiny* like quadratic forms or even *illuminating*, like projectors. And the matrices? Matrices continue to play a central role, they are the *pivot* of the situation. They are *linear objects* but they are even well adapted to model equations of the second degree. Tied to the concept of orthogonal projection are the so-called *projectors*, which give us the essential tool needed for the solution of the famous *problem of least squares*. To reach this goal we will need the help of mathematical concepts which are a bit more sophisticated, such as *vector spaces* with their systems of generators and their bases, if possible orthogonal and even better *orthonormal* and the notion of the *pseudoinverse* of a matrix.

We will also highlight the symmetric matrices. Why? Well, some claim that mathematicians choose the objects they want to study by using aesthetic criteria. And indeed, symmetry is an aesthetic choice. But, it more often happens that certain properties which appear to be *only* aesthetic are absolutely crucial for practical applications. This is the case for the symmetric matrices, which are the *soul* of quadratic forms. Around these objects (and not only for them) we will, at the end of the book, develop themes and concepts such as *eigenvalues, eigenvectors* and *invariant subspaces*. Although these objects have some strange sounding names, they are of great use and that will begin to become clear towards the end of the book.

As I said earlier, in the second part of the book we continue to emphasize concepts and examples, but not proofs and not the formal aspects of the subject.

as I said earlier, I never repeat myself

And if someone wants to go further? No problem. This is one of the intentions of the book. But, there is a certain warning that comes with this. It's enough to wander around in a mathematics library or navigate the ocean that is the Internet to find an impressive quantity of material. In fact, as I said earlier (and at the risk of repeating myself), linear algebra is one of the basic underpinnings of science and technology and thus has stimulated, and continues to stimulate, many authors. Consequently, the going gets a lot tougher and is certainly not *for everyone*.

We now turn to some aspects of style in the book, in particular on our choices with the notation. In the Italian tradition, decimal numbers are written using commas as separators, for example 1, 26 (one comma twenty six) with the period being reserved as a separator for very large numbers, e.g. 33.200.000 (thirty-three million two hundred thousand). In the Anglo-Saxon tradition one does the opposite, and thus $2, 200.25 means two thousand two hundred dollars and twenty-five cents. What should we use? The impulse of *nationalistic pride* should make us opt for the first solution. But, the fact is that our lives are lived in contact with mechanical calculators and thus conditioned by software protocols which use English as the base language. The choice thus falls on the second method. Thus, when there are strong practical or aesthetical reasons to use a separator we will write, for example, 1.26 to say 'one unit and 26 hundreths' and we will write $34,200$ to say thirty-four thousand two hundred.

Another obvious aspect is the presence in the book of self-referential phrases, aphorisms, jokes, citations and palindromes, and the reader will be immediately aware that in many cases, they are written *on the right side of the page*, beginning with *lower case letters* and *finishing without punctuation*. Why? The author believes that even a mathematics book should furnish signs and indications and not only technical information. These phrases are like *falling stars* which appear out of the blue and immediately disappear, leaving only an incomplete sign or sensation that the reader is encouraged to ponder.

And now I'll conclude with a warning. The book continually seeks to involve the reader in the discussion. There are frequent phrases of the type
– the reader will have to be satisfied with only a partial response...
– it won't be difficult for the reader to interpret the significance of...
I hope the female readers are not offended by my use of the pronoun "he". The choice is not an anti-feminist statement! In fact, it was only dictated by my desire not to make the text too ponderous. Let me be clear then, the reader is, for me, *he or she who is reading this book.* Finally, to *really conclude,* enjoy yourself with the following little problem and then "Good reading *to everyone!*"

little problem: complete the following sequence with the two missing symbols
                    ottffsse...

Genova, 9 October 2006   Lorenzo Robbiano
Genova, 8 July 2010   Anthony V. Geramita

Contents

Part III

in this world there are two types of people,
those who think that mathematics is useless
and those who think

Numerical and Symbolic Computations

two thirds of Italians don't understand fractions
the other half are not interested

Suppose that a reader, perhaps intrigued by the title, wanted to see immediately if the book was really *for everyone* and consequently arrived here without having read neither the Forward nor the Introduction. I believe that such a reader would have made an error and missed an essential aspect of the spirit of the book. I would strongly advise such a reader to return and read those parts of the book. However, since the reader is at liberty to do as he or she chooses, and since I personally know many readers who have the habit (may I say the *bad habit*?) of not reading introductions, I have decided not to mislead even those readers and so I will begin with this very short chapter, a typical Chapter 0, in which one does some simple computations and then discusses the results obtained by those computations.

Now, even if the title *numerical* and *symbolic calculations* is very high sounding, in fact we deal here with some questions addressed at the high school level. What do we mean when we write $ax = b$? How does one manipulate the expression $ax = b$? What does it mean to solve the equation $ax = b$?

If the reader thinks that we are dealing with trivialities it would be good to pay attention nonetheless because sometimes under apparently calm waters there move dangerous under currents; underestimating the significance of these questions could be fatal. Not only that, but reading now with maximum concentration will be very useful in giving the reader the confidence to deal with important concepts that will be fundamental in what follows. In addition, *computations, numbers and symbols* are the basic ingredients of mathematics and the reader, even if he or she doesn't aspire to become a professional mathematician, will do well to familiarize themselves with these ideas.

Robbiano L.: Linear Algebra for everyone
© Springer-Verlag Italia 2011

The equation $ax = b$. Let's try to solve it

In elementary school we learn that the division of the number 6 by the number 2 gives the *exact* answer the number 3. One can describe this fact mathematically in various ways, for example writing $\frac{6}{2} = 3$ or $6 : 2 = 3$ or saying that 3 is the *solution of the equation* $2x = 6$ or that 3 is the solution of the equation $2x - 6 = 0$.

Let's recall a few things. The first thing is that the expression $2x$ means $2 \times x$ because of the convention of not writing the symbol for the product when it is not strictly necessary. The second thing is that the expression $2x = 6$ contains the symbol x, which represents the **unknown** of the problem, or put another way, the number which when multiplied by 2 gives us 6 as a result, and also contains the two **natural numbers** 2, 6.

We observe that the solution 3 is also a natural number but that this is not always the case. It would be enough to consider the problem of dividing the number 7 by the number 4. This problem's mathematical description is the same as that above, in other words one tries to resolve the equation $4x = 7$, but this time we notice that *there does not exist a natural number* which when multiplied by 4 gives 7 as a result. At this point there are two directions in which we can go.

The first is that of using the **algorithm** we learned as children for dividing two natural numbers. This direction brings us to the solution 1.75, a so-called **decimal number**. The second direction is that of *inventing a larger place, namely that of the* **rational numbers**. Taking this second direction we arrive at the solution $\frac{7}{4}$. We observe that 1.75 and $\frac{7}{4}$ are two different representations of the same mathematical object, namely the solution to the equation $4x = 7$.

But, the situation can be much more complicated. Let's try to solve a very similar problem, namely $3x = 4$. While the solution $\frac{4}{3}$ can be found easily and quickly in the rational numbers, if we try to use the division algorithm we enter into an *infinite cycle*. The algorithm produces the number $1.3333333\ldots$ and one notes that the symbol 3 is repeated infinitely often, since at each iteration of the algorithm we find ourselves in exactly the same situation as before. We can, for example, finish by saying that the symbol 3 is *periodic* and write (making it a convention) the result as $1.\overline{3}$ or as $1.(3)$. Another way to take care of this situation is simply to exit from the cycle after a fixed number of times (say five). In that case we conclude saying that the solution is 1.33333. There is, however, a big problem. If we transform the number 1.33333 into a rational number we find $\frac{133333}{100000}$, which *is not equal to* $\frac{4}{3}$. In fact, one has

$$\frac{4}{3} - \frac{133333}{100000} = \frac{4 \times 100000 - 3 \times 133333}{300000} = \frac{400000 - 399999}{300000} = \frac{1}{300000}$$

and although $\frac{1}{300000}$ is a *very small number*, it is not zero.

Although it might be useful to work with numbers that have a *fixed number of places after the decimal* one pays the price that the result is not always exact. Why then don't we always work with *exact numbers*, for example with rational numbers?

For the moment the reader will have to be content with a partial answer, but one which suggests the essence of the problem.

- One reason is that working with rational numbers is *very costly from the point of view of computation*.
- Another reason is that *we don't always have rational numbers at our disposal* as they were in our problems above.

As for the first reason, it's enough to think about the difficulty that a calculator has in recognizing the fact that the following equivalent *fractions*, $\frac{4}{6}, \frac{6}{9}, \frac{2}{3}$, represent the *same rational number*.

As for the second reason, suppose (for example) that we wanted to find the relationship between the distance from the earth to the sun and the distance from the earth to the moon. Calling b the first distance and a the second distance, the equation that represents our problem is our old friend $ax = b$. But, no-one would contend that it was reasonable to have *exact numbers* to represent such distances. The initial data of our problem are *necessarily approximate numbers*. In this case we would consider such a difficulty as impossible to eliminate and we would take the appropriate precautions.

The equation $ax = b$. Be careful of mistakes

Let's return to our equation $ax = b$. In terms of the problem giving rise to the numbers a and b, let's think about whether we want **exact solutions** or **approximate solutions**. As we saw above, $\frac{4}{3}$ is an exact solution of $3x = 4$, or equivalently, of $3x - 4 = 0$, while 1.33333 is an approximate solution which differs from the exact solution only by $\frac{1}{300000}$ or, using another very common notation, by $3.\overline{3} \cdot 10^{-6}$. Taking also into account the fact we saw in the preceding section, namely that it is not always possible to operate with exact numbers, one begins to think that a tolerably small error is not so bad. But, real life is full of obstacles.

Suppose that our initial data were $a = \frac{1}{300000}$, $b = 1$. The correct solution is $x = 300000$. If we made an error in the initial valuation of a and said $a = \frac{2}{300000}$, then our error was only $\frac{1}{300000}$, something we recently declared to be a "tolerably small error". But now our equation $ax = b$ has solution $x = 150000$, which differs from the correct solution by 150000.

What happened? Simply put, if one divides a number b by a *very small* number a, the result is *very big*; thus if one alters the number a by a very small amount, the result is changed by a very large amount. This problem, which we have to keep in front of us all the time when we work with approximate quantities, has given rise to a large sector of mathematics which is called **numerical analysis**.

The equation $ax = b$. Let's manipulate the symbols

The discussion we have just had about the quantities a and b and about approximate solutions has nothing to do with purely formal, or symbolic, manipulations. For example, little kids learn that starting with the equation $ax = b$, one can write an equivalent equation by *moving b to the left of the equality and changing its sign*. This is an example of a **symbolic computation**, more precisely of the use of a *rewrite rule*.

Just exactly what does that mean? If α is a solution of our equation then we have the equality of numbers $a\alpha = b$ and thus the equality $a\alpha - b = 0$. This observation permits us to conclude that the equation $ax = b$ is *equivalent* to the equation $ax - b = 0$, in the sense that they have the *same solutions*. The transformation of $ax = b$ into $ax - b = 0$ is a manipulation that is purely symbolic, *independent of the nature of the problem*. This would be a good time to comment on the fact that such a manipulation is *not always* valid. If, for example, we are working with natural numbers, the expression $2x = 4$ cannot be transformed into $-4 + 2x = 0$, since -4 is not a natural number.

Now I would like us to take a leap and *solve* the equation *independent of the values of a and b*. In other words, we would like to find an expression for the solution of $ax = b$ (or equivalently $ax - b = 0$) which depends only on a and b and not on any particular values that we might attribute to them.

Put in that generality, it's not really possible. For example, what happens if $a = 0$? In that situation there are two possible cases, depending on whether $b \neq 0$ or $b = 0$. In the first case we can say definitely *there are no solutions* because there is no number which when multiplied by zero produces a number that is different from zero. In the second case, instead, *all numbers are solutions* because every number when multiplied by zero gives zero.

It thus seems that when $a = 0$ the equation $ax = b$ behaves in two extremely different ways. The situation becomes more manageable if we suppose that $a \neq 0$; in that case we can immediately conclude that $\frac{b}{a}$ is the unique solution. But, can we be sure? Didn't we already say in the preceding section that the equation $4x = 7$ *does not have integer solutions*? Certainly 4 is different from 0!

The problem is the following. In order to conclude that if $a \neq 0$, then $\frac{b}{a}$ is the solution to $ax = b$, we have to know that $\frac{b}{a}$ *makes sense*. Without entering into all the algebraic refinements that this questions implies, we'll limit ourselves to observing that the rational numbers, the real numbers and the complex numbers have the property that if a is a rational, real or complex number different from zero, then it has an *inverse* (which in algebra one calls a^{-1}). For example, the inverse of 2 in the rational numbers is $\frac{1}{2}$, while in the whole numbers the inverse doesn't exist.

This kind of argumentation has an *exquisitly* mathematical nature, but its importance for applications is revealing itself to be of increasing importance. Current technology actually puts at our disposal hardware and software with which we can manipulate data symbolically. There is a new area of mathematics concerned with these things; and it is emerging as a strong area of study. It is known as *symbolic computation* but also called **computational algebra** or **computer algebra** (see [R06]).

Exercises

Before you begin to consider the problems given in the exercises, permit me to offer some advice. The reader should remember that, as well as the techniques learned in each section, it is always of fundamental importance to use common sense when approaching a problem. I'm not kidding! In fact, it often happens that university students concentrate so hard on trying to use the formulas they have learned in the course, that they don't realize that a small dose of common sense is often what is needed to solve the problems. Even if that common sense is not enough, it will (in any case) help.

Exercise 1. What power of 10 is a solution to $0.0001x = 1000$?

Exercise 2. Consider the equation $ax - b = 0$, where $a = 0.0001$, $b = 5$.

(a) Find the solution α.

(b) By how much do you have to alter a in order to have a solution that differs from α by at least 50000?

(c) If p is a positive number that is smaller than a, can you produce a bigger error by substituting for a the number $a - p$ or the number $a + p$?

Exercise 3. Construct an example of an equation of type $ax = b$, in which an error in the coefficients hardly makes any difference in the error of the solution.

Exercise 4. Despite the fact that the inverse of 2 doesn't exist in the integers, why is it possible to solve (with integers) the equation $2x - 6 = 0$?

Exercise 5. Are the two equations $ax - b = 0$ and $(a-1)x - (b-x) = 0$ equivalent?

Exercise 6. Consider the following equations (with a parameter) of type $ax - b = 0$.

(a) Find the real solutions of $(t^2 - 2)x - 1 = 0$ in terms of t in $\mathbb{Q}$.

(b) Find the real solutions of $(t^2 - 2)x - 1 = 0$ in terms of t in $\mathbb{R}$.

(c) Find the real solutions of $(t^2 - 1)x - t + 1 = 0$ in terms of t in $\mathbb{N}$.

(d) Find the real solutions of $(t^2 - 1)x - t + 1 = 0$ in terms of t in $\mathbb{R}$.

Part I

1

Systems of Linear Equations and Matrices

linear hypotheses in a non-linear world
are highly dangerous

(Adam Hamilton)

In the introductory chapter we warmed up the engines by studying the equation $ax = b$. What will we do to follow up on that? I will tell you right away that in this chapter we will deal with transportation problems and chemical reactions, diet manipulations, totocalcio tickets, architectural constructions and meteorology. How is that possible? Am I changing the scope of the book? On the contrary. The fascination of even not very sophisticated mathematics lies precisely in its ability to bring together topics which, at the outset, seem very different.

In fact, we will see many examples that will appear totally different, but we will discover that they can all be united by a single, simple mathematical model, the so-called *system of linear equations*. It will then be natural to ask how one represents a system of linear equations and at that point the *leading ladies* enter the scene, the objects that will play the primary role right up to the end, the *matrices*.

As one of the first results of our investigations, we will see how matrices allow us to use the formalism $A\mathbf{x} = \mathbf{b}$ for systems of linear equations. This looks a lot like the familiar equation $ax = b$ with which we began. I think we can all agree on this.

The more expert reader may observe, however, that the world in which we live is not, in general, linear, and that life is rife with obstacles. That's true, but if one learns to look with care, one discovers many *linear phenomena* even where one least expects them. Are you curious to know where? A little patience and you will find out, but you will have to be a collaborator in this venture. For example, you will have to learn how to deal not only with systems of linear equations and matrices but also with vectors.

Robbiano L.: Linear Algebra for everyone
© Springer-Verlag Italia 2011

1.1 Examples of Systems of Linear Equations

First let's look at some examples. The first is our good friend from the preceding chapter.

Example 1.1.1. The equation ax = b
As I said, the first example puts the equation $ax = b$ in this new context which we are calling **systems of linear equations**.

The second example will be an old friend to those who have studied analytic geometry.

Example 1.1.2. The line in the plane
The linear equation $ax + by + c = 0$ represents a line in the plane. In a little while (in fact at the end of Section 4.2) we will make clear what we mean when we say that an equation represents a geometric object.

And if you want to intersect two lines in the plane?

Example 1.1.3. Intersection of two lines

$$\begin{cases} ax +by +c = 0 \\ dx +ey +f = 0 \end{cases}$$

You'll notice that we need lots of letters! But, let's not allow this to bother us now and let's resolutely move on to see something more interesting.

Example 1.1.4. Transportation
Suppose there are two factories F_1, F_2 which produce, respectively, $(120, 204)$ automobiles. Suppose further that the factories need to move their cars to two car dealers, D_1, D_2 who want, respectively, 78 and 246 cars. Notice that in this situation we have

$$120 + 204 = 78 + 246 = 324$$

and thus we are in a position to make a transportation plan. For example, if we call x_1, x_2 the number of cars that factory F_1 will transport to dealers D_1, D_2 respectively, and y_1, y_2 the number of cars that factory F_2 will transport to the dealers D_1, D_2 respectively, then we must have

$$\begin{cases} x_1 + x_2 = 120 \\ y_1 + y_2 = 204 \\ x_1 + y_1 = 78 \\ x_2 + y_2 = 246 \end{cases}$$

This is a system of linear equations. Our transportation problem has been translated into a mathematic model; in other words this system of linear equations has captured the mathematical essence of the problem.

There are 4 equations and 4 unknowns. Can we hope that there is a solution or even many solutions? For the moment we don't have the technical tools to allow us to answer the question, however, by trial and error, we find that $x_1 = 78$, $x_2 = 42$, $y_1 = 0$, $y_2 = 204$ is a solution. Not only, but also $x_1 = 70$, $x_2 = 50$, $y_1 = 8$, $y_2 = 196$ is a solution and even $x_1 = 60$, $x_2 = 60$, $y_1 = 18$, $y_2 = 186$ is a solution. It seems clear that there are many solutions. How many? And why is it important to know all of the solutions?

We may suppose that the unit cost of transporting the cars from the factories to the dealers are not the same, for example, suppose the cost of transporting a car from factory F_1 to dealer D_1 is 10 Euros, from factory F_2 to dealer D_2 is 9 Euros, from F_2 to D_1 is 13 Euros and from F_2 to D_2 is 14 Euros. The three solutions we noted above would then have a total cost respectively of

$$10 \times 78 + 9 \times 42 + 13 \times 0 + 14 \times 204 = 4,014 \text{ Euro}$$

$$10 \times 70 + 9 \times 50 + 13 \times 8 + 14 \times 196 = 3,998 \text{ Euro}$$

$$10 \times 60 + 9 \times 60 + 13 \times 18 + 14 \times 186 = 3,978 \text{ Euro}$$

The third solution thus is the *cheapest*. But, among all the possible solutions is that one really the cheapest? In order to answer that question (and others like it) it seems clear that we really need to know *all the solutions* and we will shortly be able to find them all. In particular we will understand that the answer to the question we just posed (i.e. is it really the cheapest?) will be decidedly no.

Example 1.1.5. A chemical reaction

Combining atoms of *copper* (Cu) with molecules of *sulfuric acid* (H_2SO_4), gives rise to molecules of *copper sulphate* ($CuSO_4$), water (H_2O) and *sulphur dioxide* (SO_2). We would like to determine the number of molecules which take part in this chemical reaction. We will indicate with x_1, x_2, x_3, x_4, x_5 respectively, the number of molecules of Cu, H_2SO_4, $CuSO_4$, H_2O and SO_2. The chemical reaction is then expressed with an equality of the type

$$x_1 Cu + x_2 H_2SO_4 = x_3 CuSO_4 + x_4 H_2O + x_5 SO_2$$

In reality the equality doesn't completely describe the situation since the chemical reaction is oriented, in other words it *starts from the left and arrives at the right*. On the other hand, the constraint is that the number of atoms of every element has to be the same before and after the reaction (in *mathematese* one would say that the number of atoms of each element is an **invariant** of the chemical reaction). For example, the number of atoms of oxygen (O) is $4x_2$ from the left side of the equality and $4x_3 + x_4 + 2x_5$ from the right side and so we must have $4x_2 = 4x_3 + x_4 + 2x_5$.

Thus, the five numbers x_1, x_2, x_3, x_4, x_5 are constrained by the following relations

$$\begin{cases} x_1 & - & x_3 & & & = 0 \\ & 2x_2 & & - & 2x_4 & & = 0 \\ & x_2 & - & x_3 & & - & x_5 = 0 \\ & 4x_2 & - & 4x_3 & - & x_4 & - & 2x_5 = 0 \end{cases}$$

This is a system of linear equations. It is a mathematical model of the above mentioned chemical problem. In other words we have captured the essence of the mathematical problem posed by the invariance of the number of atoms of each element present both before and after the reaction (i.e. on the left side and on the right side of the equation).

The natural next step is to solve the system of equations. We've not yet seen how to do that, but in this case we can try to make some *experimental calculations*. The first equation tell us that $x_1 = x_3$ and the second that $x_2 = x_4$. Thus, in the third and fourth equations we can substitute x_3, whenever there is an x_1 and an x_4 whenever there is an x_2. In this way we get $x_4 - x_3 - x_5 = 0$ and $3x_4 - 4x_3 - 2x_5 = 0$. From the first of these we see that $x_3 = x_4 - x_5$ and we use that to substitute for x_3 in the second and thus obtain $-x_4 + 2x_5 = 0$, and hence the equality $x_4 = 2x_5$.

Going back to our previous substitutions we get $x_3 = x_5$ and so, reconsidering the first two equations we get $x_1 = x_5$, $x_2 = 2x_5$.

This discussion has, for the moment, a very empirical character and we will see later how to make it more rigorous. For now, let's be satisfied to note that the solutions to our system may be written in the form $(x_5, 2x_5, x_5, 2x_5, x_5)$ with x_5 being totally arbitrary. We thus have a case in which there are an infinite number of solutions except that the only ones which interest us are those for which the solutions are natural numbers (it wouldn't make sense to speak of -2 molecules) and we would like the solution to be as small as possible. This last request is similar to that already noted in the transportation problem. In our case it is easy to see that there is such a solution and it is $(1, 2, 1, 2, 1)$. In conclusion, the proper chemical reaction is the following

$$Cu + 2H_2SO_4 = CuSO_4 + 2H_2O + SO_2$$

which should be read in the following way: a one atom molecule of copper and two molecules of sulfuric acid react chemically to give rise to a molecule of copper sulfate, two molecules of water and one molecule of sulfur dioxide.

Example 1.1.6. The diet

Suppose we want to prepare a breakfast with butter, ham and bread in such a way as to have available 500 calories, 10 grams of protein and 30 grams of fat. A chart shows the number of calories, the amount of protein (expressed in grams) and the amount of fat (expressed in grams) furnished by a gram of butter or ham or bread.

	butter	ham	bread
calories	7.16	3.44	2.60
protein	0.006	0.152	0.085
fat	0.81	0.31	0.02

If we denote by x_1, x_2, x_3 respectively, the number of grams of butter, ham and bread, then what we are looking for is nothing more than the solution to the system of linear equations

$$\begin{cases} 7.16\,x_1 + 3.44\,x_2 + 2.60\,x_3 = 500 \\ 0.006\,x_1 + 0.152\,x_2 + 0.085\,x_3 = 10 \\ 0.81\,x_1 + 0.31\,x_2 + 0.02\,x_3 = 30 \end{cases}$$

Again we will have to be a bit patient with this problem. We will see, in a bit, how we can resolve this system. In the meantime, probably best not to be tempted into eating too much.

Example 1.1.7. The bridge

A bridge is to be built to connect the two sides of a river which are at different levels, as indicated in the figure. One assumes that the profile of the bridge is parabolic and that the parameters indicated by the architect are p_1, p_2, c, ℓ, where: p_1 represents the slope of the bridge at the point A where the bridge connects to the first shore of the river, p_2 represents the slope of the bridge at the point B where the bridge connects to the other shore, c represents the height of the first shore at the point where the bridge is attached there and ℓ represents the width of the river bed underneath the bridge.

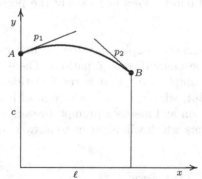

The problem is to determine the maximum height of the bridge as a function of the parameters mentioned above. In order to set up the solution we must use some notions of analytic geometry which we will assume are known.

If we fix the orthogonal cartesian axes as in the figure, the generic equation of a parabola is of the form $y = ax^2 + bx + c$. The coefficient c is exactly that which we have already called c, since it is the intersection of the parabola with the y axis and that point has coordinates $(0, c)$. The point A has abscissa 0, the point B has abscissa ℓ; the first derivative of $ax^2 + bx + c$ with respect to x is $2ax + b$ and thus has value b at the point A, and $2a\ell + b$ at the point B.

Consequently we have $p_1 = b$, $p_2 = 2a\ell + b$. In order to determine the value of a, b, and thus find the equation of the parabola, we must solve the system

$$\begin{cases} b - p_1 & = 0 \\ 2a\ell + b & - p_2 = 0 \end{cases}$$

in which a and b are the unknowns and p_1, p_2, ℓ are the parameters. Naturally, in this case the solution is easy to find and we have $b = p_1$, $a = \frac{p_2 - p_1}{2\ell}$ and the equation of the parabola is thus

$$y = \frac{p_2 - p_1}{2\ell} x^2 + p_1 x + c$$

At this point we can observe that the height of the shore at the point B, that is the ordinate of B is fixed and equal to $\frac{p_2 - p_1}{2\ell}\ell^2 + p_1\ell + c$, that is $\frac{p_1 + p_2}{2}\ell + c$.

Setting the first derivative of the equation of the parabola equal to 0 one has $2\frac{p_2 - p_1}{2\ell}x + p_1 = 0$, from which we deduce $x = \frac{\ell p_1}{p_1 - p_2}$ and thus the ordinate of the maximum point

$$y = \frac{p_2 - p_1}{2\ell}\left(\frac{\ell p_1}{p_1 - p_2}\right)^2 + p_1 \frac{\ell p_1}{p_1 - p_2} + c$$

which we can simplify to

$$y = \frac{\ell p_1^2}{2(p_1 - p_2)} + c$$

The **parametric solution** allows us to study the problem by varying the initial data.

We have thus seen many examples of systems of linear equations which can be adapted to describe many diverse situations. There is one characteristic which unites these examples, and that is the fact that all are **systems of linear equations**. But, what exactly is a system of linear equations? This is an important question and merits a prompt answer. The next section will prepare the instruments which will allow us to answer this question.

1.2 Vectors and Matrices

Two types of mathematical objects are of fundamental importance in what follows, the **vectors** and the **matrices**. Thus, before responding to the question which we have left up in the air, let's consider some significant examples.

Example 1.2.1. Velocity
The idea of vector used in physics is well known. They are used, for example, to express force, acceleration and velocity. We'll consider these examples a bit ahead, but for now it will suffice to say that a velocity vector is written as

$v = (2, -1)$ where the meaning is that its horizontal **component** is 2 units of measure, e.g. meters per second, while the vertical component is -1 unit of measure. I repeat, we will consider this concept in Chapter 4. Let's live with this situation for a while (not a long while).

Example 1.2.2. The totocalcio ticket

A completely different kind of example is afforded by the totocalcio lottery ticket. It is an example of a completely different kind of vector, which can be represented with an ordered sequence of numbers and symbols, e.g.

$$s = (1, X, X, 2, 2, 1, 1, 1, 1, X, X, 1, 1)$$

Example 1.2.3. Temperature

Let N, C, S be three Italian locations, one in the north, one in the center and one in the south. Suppose that information has been kept about the median temperature in those locations for the last 12 months. How do you imagine such information is kept? One writes

$$\begin{pmatrix} & J & F & M & A & M & J & Jul & A & S & O & N & D \\ N & 4 & 5 & 8 & 12 & 16 & 19 & 24 & 25 & 20 & 16 & 9 & 5 \\ C & 6 & 7 & 11 & 15 & 17 & 24 & 27 & 28 & 25 & 19 & 13 & 11 \\ S & 6 & 8 & 12 & 17 & 18 & 23 & 29 & 29 & 27 & 20 & 14 & 12 \end{pmatrix}$$

and then one surely has a clear view of all the available information. But, there is an *overabundance* of data, a certain lack of homogeneity in it.

If we denote the three locations with the symbols L_1, L_2, L_3 and the months of the year with the symbols $M_1, \ldots, M_{12}$, then we don't need either the first row or the first column of the matrix above. It would be enough to write

$$\begin{pmatrix} 4 & 5 & 8 & 12 & 16 & 19 & 24 & 25 & 20 & 16 & 9 & 5 \\ 6 & 7 & 11 & 15 & 17 & 24 & \mathbf{27} & 28 & 25 & 19 & 13 & 11 \\ 6 & 8 & 12 & 17 & 18 & 23 & 29 & 29 & 27 & 20 & 14 & 12 \end{pmatrix}$$

In fact, for example, the number 27 written in boldface, which we find in the *second row*, refers to the second location, L_2, i.e. to the central location and since it lies in the *seventh column* it refers to the seventh month, i.e. the month of July. The advantages are the following:

– the display is a bit smaller;
– the data in the table are homogeneous (temperature).

We can thus consider the vectors and the matrices as *containers of numerical information*. Taking another look at the preceding examples, a natural question comes to mind: are vectors only special matrices? In order to answer this, let's take another look at the Example 1.2.2. We've indicated with $s = (1, X, X, 2, 2, 1, 1, 1, 1, X, X, 1, 1)$ a particular totocalcio ticket, put another way, we have used the vector s to contain the thirteen symbols $1, X, X, 2, 2, 1, 1, 1, 1, X, X, 1, 1$. Naturally we can consider s as a *matrix with only one row*, but someone might say that usually in a totocalcio ticket the symbols are written in a single column

$$\begin{pmatrix} 1 \\ X \\ X \\ 2 \\ 2 \\ 1 \\ 1 \\ 1 \\ 1 \\ X \\ X \\ 1 \\ 1 \end{pmatrix}$$

We'll close this discussion, for the moment, noting that a vector may be seen both as a row matrix or as a column matrix because the information we are encoding is the same. It's useful to know that the usual *convention* is that *vectors are viewed as column vectors*. Nevertheless, it is worthwhile to clearly state that in mathematics vectors and matrices are considered as *different entities* and then when one speaks of row vectors or column vectors, one is really speaking of a row matrix or a column matrix that *represents* the vector, but they are not the same thing as the vector. Nevertheless, in what follows we will often make the identification mentioned above.

1.3 Generic Systems of Linear Equations and Associated Matrices

We are finally ready to answer the question left up in the air at the end of Section 1.1. Recall that the question was: just what is a system of linear equations? Let's consider, for simplicity, the case of a single equation, for example $x - 3y = 0$. What do we notice? The principal feature is that the *unknowns* x, y appear in the expression with exponent one. But that's not all, in fact it's also the case that in the expression $xy - 1$ and even in the expression $e^x - 1$, the unknowns appear with exponent one. We have to be a bit more technically precise and the exact definition is the following:

> *A linear expression is a polynomial expression in which all the monomials are of degree less than or equal to one.*

That description excludes the expression $e^x - 1$, which is not a polynomial and also excludes the expression $xy - 1$ because the monomial xy has degree 2.

An example of a **linear equation** is $x - \frac{2}{3}y - 1 = 0$ and the generic linear equation with one unknown is our old friend $ax - b = 0$. Instead of the word *unknown* algebraists often use the word **indeterminate**.

Let's look at an example that *everyone* knows. In a square with side ℓ, the perimeter p is given by the expression $p = 4\ell$, while the area A is given by the expression $A = \ell^2$. We observe immediately that ℓ^2 is not linear, since the exponent of ℓ is 2. In fact, such an expression is called *quadratic*. The practical consequences are very obvious: if, for example, we double the side, the perimeter is doubled (linear effect) while the area is quadrupled (quadratic effect).

How does one make a **generic system of linear equations**? But, before we start, we should think about the meaning of the word *generic*. We have already observed that in order to deal with problems, even of very different types, it is easy to find oneself with a system of linear equations. These systems of equations represent that which is commonly called a **mathematical model** of the problem. And even if the problems are completely different, once their model is that of a system of linear equations, they can all be treated in the same way. This is the real strength of mathematics!

But, in order to unify the treatment it is opportune to find the right language. For example, the first problem to deal with is that of deciding how to write a generic system of linear equations so that every system of linear equations can be seen as a *particular case* of that system. We will have to use some symbols, both for the number of rows and for the number of unknowns, and also for the coefficients and the unknowns. In the same way that we used $ax = b$ to describe a generic linear equation with one unknown, we will (using both letters and indices) find a way to write the generic system of linear equations, which for convenience we will denote by $\mathcal{S}$. Let's see how we describe $\mathcal{S}$.

$$\begin{cases} a_{11}x_1 + a_{12}x_2 + \cdots + a_{1c}x_c & = & b_1 \\ a_{21}x_1 + a_{22}x_2 + \cdots + a_{2c}x_c & = & b_2 \\ \cdots\cdots\cdots\cdots\cdots\cdots\cdots\cdots\cdots & = & \cdots \\ a_{r1}x_1 + a_{r2}x_2 + \cdots + a_{rc}x_c & = & b_r \end{cases}$$

At first glance this all looks a bit mysterious, but let's try to analyze it attentively. Let's start with the question: what does r mean? It's the name given to the number of rows, in other words to the number of equations. What about c? It's the name given to the number of columns, in other words to the number of unknowns. Why have we put a double index alongside the coefficients? The reason is that this artifice allows us to identify the coefficient in an unambiguous way. For example a_{12} is the name of the coefficient of x_2 in the first equation, a_{r1} is the name of the coefficient of x_1 in the r-th equation, and so on. Inasmuch as we are at a rather abstract level, let's use that as an excuse to mention that mathematicians use the name **homogeneous system of linear equations** for such a system in which all the **constant terms** $b_1, b_2, \ldots, b_r$ are zero.

Let's try to see if we have actually succeeded in describing *every* system of linear equations. Let's consider, for example, the following system of linear equations having two equations and four unknowns

$$\begin{cases} 5x_1 + 2x_2 - \frac{1}{2}x_3 - \quad x_4 = 0 \\ x_1 \qquad\quad - \ x_3 + \frac{12}{5}x_4 = 9 \end{cases}$$

Let's try to identify it as a particular case of S. We immediately see that $a_{11} = 5$, $a_{12} = 2$, $a_{13} = -\frac{1}{2}$, $a_{14} = -1$, $b_1 = 0$ and so on. But, what's happened to a_{22}? Naturally, $a_{22} = 0$ and thus we didn't write the term $0x_2$ in the second equation. Notice that, in this example, we have $r = 2$ and $c = 4$.

Let's go back and look at the examples in Section 1.1. We will see that one does not always use the general scheme S because there might be special reasons not to do so. In Example 1.1.2, instead of writing $a_{11}x_1 + a_{12}x_2 = b_1$, we wrote $ax + by + c = 0$. That seems very different, but let's look closely at the differences.

The most noticeable thing is that we haven't used the double indices. Why not? The reason should be clear. If the system consists of only one single equation it's not necessary to use an index to indicate the *unique row*. But, in fact, we haven't even used the column index since that wasn't necessary either. Since we had to name only three coefficients we choose to use a, b and c instead of a_1, a_2 and a_3. Finally we wrote $ax + by + c = 0$ instead of $ax + by = -c$, but the fact that we could do this was already discussed in the introductory chapter.

In Example 1.1.4 we used a different strategy for naming the unknowns. If you recall, we used x_1 and x_2 to indicate the number of cars made by factory F_1 that were sent, respectively, to car dealers D_1, D_2 and by y_1, y_2, the number of cars made by factory F_2 and sent to, respectively, the car dealers D_1, D_2. If we wanted to generalize this example, the non-standard choice that we made above would create some difficulties and, in fact, Exercise 5 is there precisely so that you can appreciate this observation.

Let's return to the fundamental problem of identifying a system of linear equations, and thus let's go back to the generic system S

$$\begin{cases} a_{11}x_1 + a_{12}x_2 + \cdots + a_{1c}x_c &= b_1 \\ a_{21}x_1 + a_{22}x_2 + \cdots + a_{2c}x_c &= b_2 \\ \quad\cdots\cdots\cdots\cdots\cdots\cdots\cdots\cdots &= \cdots \\ a_{r1}x_1 + a_{r2}x_2 + \cdots + a_{rc}x_c &= b_r \end{cases}$$

Once we have decided to use r to denote the number of equations, c to denote the number of unknowns and, moreover, to call the unknowns of the system $x_1, x_2, \ldots, x_c$ we realize that we have done something very suggestive. For example, we could have called m the number of equations, n the number of unknowns and $y_1, y_2, \ldots, y_n$ the unknowns themselves. Having done that we would not have changed anything (apart from the *physical appearance* of the system itself). What then really characterizes the system S?

The answer is that the **characterizing aspects of the system** S are the coefficients a_{ij} and b_j, as we vary i from 1 to r and j from 1 to c. To make this concept clearer let's consider the following two systems of linear equations

$$\begin{cases} x_1 + x_2 - x_3 = 0 \\ x_1 - 2x_2 + \frac{1}{4}x_3 = \frac{1}{2} \end{cases} \qquad \begin{cases} y_1 + y_2 - y_3 = 0 \\ y_1 - 2y_2 + \frac{1}{4}y_3 = \frac{1}{2} \end{cases}$$

In this case we are really looking at the same system written in two different ways. Now let's consider the following two systems of linear equations

$$\begin{cases} x_1 + x_2 - x_3 = 0 \\ x_1 - 2x_2 + \frac{1}{4}x_3 = \frac{1}{2} \end{cases} \qquad \begin{cases} x_1 + 2x_2 - x_3 = 0 \\ x_1 - 2x_2 + \frac{1}{4}x_3 = 1 \end{cases}$$

In this case we are really looking at two different systems, even if we have used the same names for the unknowns.

It starts to become clear that the information in the system S is totally contained in the **coefficient matrix**

$$A = \begin{pmatrix} a_{11} & a_{12} & \cdots & a_{1c} \\ a_{21} & a_{22} & \cdots & a_{2c} \\ \vdots & \vdots & \vdots & \vdots \\ a_{r1} & a_{r2} & \cdots & a_{rc} \end{pmatrix}$$

and in the vector of constant terms

$$\mathbf{b} = \begin{pmatrix} b_1 \\ b_2 \\ \vdots \\ b_r \end{pmatrix}$$

or, if you prefer, in the **augmented matrix**

$$B = \begin{pmatrix} a_{11} & a_{12} & \cdots & a_{1c} & b_1 \\ a_{21} & a_{22} & \cdots & a_{2c} & b_2 \\ \vdots & \vdots & \vdots & \vdots & \vdots \\ a_{r1} & a_{r2} & \cdots & a_{rc} & b_r \end{pmatrix}$$

We have thus discovered the following fact.

The information of a system of linear equations can be completely expressed with the use of matrices and vectors.

Since from now on we will be using the language of matrices quite a bit, it would be good to establish some conventions about how (and what) we will call things. For example, the elements that appear in a matrix are called the **entries**. So $a_{12}, b_1, \ldots$ are entries of the matrix B.

At this point it should be clear that when we speak of a generic matrix with r rows and c columns we mean the following

$$A = \begin{pmatrix} a_{11} & a_{12} & \cdots & a_{1c} \\ a_{21} & a_{22} & \cdots & a_{2c} \\ \vdots & \vdots & \vdots & \vdots \\ a_{r1} & a_{r2} & \cdots & a_{rc} \end{pmatrix}$$

We say that

- *the matrix A is of* **type** (r, c), *to indicate that it has r rows and c columns;*
- *the matrix A is* **square** *of type r to say that the matrix has both r rows and r columns.*

Another way to represent a generic matrix is the following

$$A = (a_{ij}), \ i = 1, \ldots, r, \ j = 1, \ldots, c$$

which captures, in symbols, the following meaning

> *The matrix A has as generic entry the number a_{ij}, which has a variable double index. The row index i varies from 1 to r while the column index j varies from 1 to c.*

Notice that having written the generic matrix in this way we have not been precise about the exact nature of the entries. If we want to specify, for example, that the entries are rational numbers, then we can write

$$A = \begin{pmatrix} a_{11} & a_{12} & \cdots & a_{1c} \\ a_{21} & a_{22} & \cdots & a_{2c} \\ \vdots & \vdots & \vdots & \vdots \\ a_{r1} & a_{r2} & \cdots & a_{rc} \end{pmatrix} \qquad a_{ij} \in \mathbb{Q}$$

or

$$A = (a_{ij}), \ a_{ij} \in \mathbb{Q}, \ i = 1, \ldots, r, \ j = 1, \ldots, c$$

We conclude with a mathematical refinement. An extremely synthetic way of expressing the fact that A is a matrix with r rows, c columns and rational entries is the following

$$A \in \mathrm{Mat}_{r,c}(\mathbb{Q})$$

Notice that in order to give some significance to what we just wrote, mathematicians invented a name.

> $\mathrm{Mat}_{r,c}(\mathbb{Q})$ *is the name given to the set of all the matrices with r rows, c columns and rational entries.*

1.4 The Formalism of Ax = b

Let's reexamine, one more time, our system of linear equations $\mathcal{S}$. As we have in part already noted, the information contained in $\mathcal{S}$ can be divided into three pieces of data, of which two are the coefficient matrix A and the vector or column matrix of the constant terms $\mathbf{b}$, and the third is the vector or column matrix of the unknowns

$$\mathbf{x} = \begin{pmatrix} x_1 \\ x_2 \\ \vdots \\ x_c \end{pmatrix}$$

On seeing this, a mathematician is tempted to *imitate* the equation $ax = b$ that we discussed in the introductory chapter, and write the system $\mathcal{S}$ as $A\mathbf{x} = \mathbf{b}$. How could that be possible?

In order to make sense out of what we just wrote we have to *invent* a product $A\mathbf{x}$, that will give us the following column matrix as a result (notice that the result should have only one element in each row)

$$\begin{pmatrix} a_{11}x_1 + a_{12}x_2 + \cdots + a_{1c}x_c \\ a_{21}x_1 + a_{22}x_2 + \cdots + a_{2c}x_c \\ \dots\dots\dots\dots\dots\dots\dots \\ a_{r1}x_1 + a_{r2}x_2 + \cdots + a_{rc}x_c \end{pmatrix}$$

in order to be able to say that

$$\begin{pmatrix} a_{11}x_1 + a_{12}x_2 + \cdots + a_{1c}x_c \\ a_{21}x_1 + a_{22}x_2 + \cdots + a_{2c}x_c \\ \dots\dots\dots\dots\dots\dots\dots \\ a_{r1}x_1 + a_{r2}x_2 + \cdots + a_{rc}x_c \end{pmatrix} = \begin{pmatrix} b_1 \\ b_2 \\ \dots \\ b_r \end{pmatrix}$$

represents the same thing as

$$\begin{cases} a_{11}x_1 + a_{12}x_2 + \cdots + a_{1c}x_c &=& b_1 \\ a_{21}x_1 + a_{22}x_2 + \cdots + a_{2c}x_c &=& b_2 \\ \dots\dots\dots\dots\dots\dots\dots &=& \dots \\ a_{r1}x_1 + a_{r2}x_2 + \cdots + a_{rc}x_c &=& b_r \end{cases}$$

But now this is simple! It will be enough to *invent a product of matrices* for which

$$\begin{pmatrix} a_{11} & a_{12} & \cdots & a_{1c} \\ a_{21} & a_{22} & \cdots & a_{2c} \\ \vdots & \vdots & \vdots & \vdots \\ a_{r1} & a_{r2} & \cdots & a_{rc} \end{pmatrix} \begin{pmatrix} x_1 \\ x_2 \\ \vdots \\ x_c \end{pmatrix} = \begin{pmatrix} a_{11}x_1 + a_{12}x_2 + \cdots + a_{1c}x_c \\ a_{21}x_1 + a_{22}x_2 + \cdots + a_{2c}x_c \\ \dots\dots\dots\dots\dots\dots\dots \\ a_{r1}x_1 + a_{r2}x_2 + \cdots + a_{rc}x_c \end{pmatrix}$$

Said in this way it seems a purely artificial thing, but instead we are dealing with a point of fundamental importance. In the next chapter, in particular in

Section 2.2 we will study, and put in the proper context, the concept of a row by column product of two matrices. But even now we can use the formalism $A\mathbf{x} = \mathbf{b}$ to describe a generic system of linear equations. The student should begin to get the idea that this is not just a mere formalism but that $A\mathbf{x}$ really represents a product, i.e. a row by column product of A and $\mathbf{x}$, as we will see in detail in just a little bit.

Exercises

Exercise 1. What do the following two matrices have *in common*?

$$A = \begin{pmatrix} 1 \\ 2 \end{pmatrix} \qquad B = (1 \quad 2)$$

Exercise 2. Consider the system of linear equations

$$\begin{cases} x_1 + 2x_2 \quad -\frac{1}{2}x_3 = 0 \\ \qquad -x_2 + 0.02x_3 = 0.2 \end{cases}$$

Describe r, c, a_{21}, b_2.

Exercise 3. How does one write a generic matrix with two rows and three columns?

Exercise 4. Is it true that for the coefficient matrix B, which we discussed in Section 1.3, the following formula is valid: $B \in \mathrm{Mat}_{r,c+1}(\mathbb{Q})$?

Exercise 5. Consider the Example 1.1.4 and generalize it by substituting four letters for the numbers $120, 204$ and $78, 246$. Given that we used F_1, F_2, D_1, D_2 respectively to represent the factories and the dealers, decide which of the representations proposed below seem most relevant

$$\begin{cases} x_1 + x_2 &= f_1 \\ y_1 + y_2 &= f_2 \\ x_1 + y_1 &= r_1 \\ x_2 + y_2 &= r_2 \end{cases} \qquad \begin{cases} x_1 + x_2 &= b_1 \\ y_1 + y_2 &= b_2 \\ x_1 + y_1 &= b_3 \\ x_2 + y_2 &= b_4 \end{cases} \qquad \begin{cases} x_1 + x_2 &= a \\ y_1 + y_2 &= b \\ x_1 + y_1 &= c \\ x_2 + y_2 &= d \end{cases}$$

Exercise 6. Verify that

$$\begin{pmatrix} 1 & 1 & 0 & 0 \\ 0 & 0 & 1 & 1 \\ 1 & 0 & 1 & 0 \\ 0 & 1 & 0 & 1 \end{pmatrix} \qquad \begin{pmatrix} 1 & 1 & 0 & 0 & 120 \\ 0 & 0 & 1 & 1 & 204 \\ 1 & 0 & 1 & 0 & 78 \\ 0 & 1 & 0 & 1 & 246 \end{pmatrix}$$

are, respectively, the coefficient matrix and augmented matrix of the system of linear equations associated to Example 1.1.4.

2

Operations with Matrices

it is of basic importance to attentively read the text
to check if any   has been forgotten

Matrix, matrices... how many times have we used these words. It probably won't surprise you that we will continue to use those words frequently. The matrix is one of the most useful mathematical objects we have at our disposal, a basic tool for those who use mathematics. This is the case for lots of good reasons, some of which we have already seen and some we will see shortly.

In this chapter we will look in more depth at matrices and study some manipulations it is useful to carry out with them. We will discover the importance of the matrix product called the *row by column product*. Having done that we will meet some strange objects called *graphs* and *weighted graphs*. We'll make a short *genovese*[1] detour which will tell us how much it costs to multiply two matrices.

We will also see symmetric and diagonal matrices. These play a very important role in what follows and we will discover (perhaps with sadness, perhaps with indifference) that matrix multiplication is not commutative. Then, almost by accident, we will meet some strange numerical entities, in one of which we will see that the equality $1 + 1 = 0$ holds. At that point some readers may think that, in spite of our promises, even this book is destined to lose contact with reality. Some others might ask, what use is it to have a situation in which the equality $1 + 1 = 0$ holds?

What can I say? I can reassure the reader that, if he will have the patience to wait until the end of the chapter... she will be illuminated!! Not in the Zen sense, even if it is always best to not put limits on the power of illumination, but in the sense of resolving a practical problem related to electrical circuits. In the meantime, before we arrive at this problem we will visit some wine producers, some mountain villages, some information networks and other

---

[1]The Genovese, like the Scots, are known to seek a good price when in the marketplace.

Robbiano L.: Linear Algebra for everyone
© Springer-Verlag Italia 2011

pleasant places, and then we will arrive at the illumination with the help of
the inverse of a matrix. Matrices, are we back to them again? you say. But I
have never forgotten them.

2.1 Sum and the product by a number

Let's begin with a very simple example.

Example 2.1.1. Wine producer

Let's suppose that we are trying to record, in a matrix, information about the
sales of five different wines, in a given half year period, to certain wine sellers
in three different cities. If we use the convention that the rows correspond
to the cities and the columns to the type of wine then the matrix will be of
type $(3, 5)$. We have a matrix for each half year period, hence in a given year
we will have two matrices which we will call A_1 and A_2. What is the matrix
which contains all the sales data for the whole year? Suppose that

$$A_1 = \begin{pmatrix} 120 & 50 & 28 & 12 & 0 \\ 160 & 55 & 33 & 12 & 4 \\ 12 & 40 & 10 & 10 & 2 \end{pmatrix} \quad A_2 = \begin{pmatrix} 125 & 58 & 28 & 10 & 1 \\ 160 & 50 & 30 & 13 & 6 \\ 12 & 42 & 9 & 12 & 1 \end{pmatrix}$$

It's clear that the matrix which contains the data for the whole year is obtained by summing the corresponding entries of the two matrices, in other
words

$$\begin{pmatrix} 245 & 108 & 56 & 22 & 1 \\ 320 & 105 & 63 & 25 & 10 \\ 24 & 82 & 19 & 22 & 3 \end{pmatrix}$$

Situations like this are rather common and they encourage us to *define the
sum of two matrices of the same type as that matrix of the same type which
has as its entries the sum of the corresponding entries of the two matrices.*

If we want to express such a rule in a formal way we can say: given two
matrices $A = (a_{ij})$ and $B = (b_{ij})$ of the same type, then

$$A + B = (a_{ij} + b_{ij})$$

This definition shows that $A + B = B + A$, or put another way, that the
following property holds:

The sum of matrices is commutative

If $b_{ij} = 0$ for every i, j then the matrix B is called the **zero matrix** and
it has the property that $A + B = A$. Thus, the zero matrix of a given type
behaves, with respect to the sum of matrices, in the same way that 0 behaves
with respect to the sum of numbers. The analogy is so strong that we often
call the zero matrix 0. You will have to understand from the context the
type of the zero matrix we are considering. For example, if A is a matrix of
type $(2, 3)$, then in the formula $A + 0 = A$ we realize that the 0 represents a
matrix also of size $(2, 3)$, i.e. in this context $0 = \begin{pmatrix} 0 & 0 & 0 \\ 0 & 0 & 0 \end{pmatrix}$.

Example 2.1.2. Consumer prices
Suppose we have a matrix which represents the prices of some common items in different cities. Suppose that C_1, C_2, C_3 are three cities and B_1, B_2, B_3, B_4 are the average costs of four common items in a fixed month. For example, the matrix

$$A = \begin{pmatrix} 50 & 12.4 & 8 & 6.1 \\ 52 & 13 & 8.5 & 6.3 \\ 49.3 & 12.5 & 7.9 & 6 \end{pmatrix}$$

could represent the data for the month of August in 2006. If one expects an increase in price due to inflation at the rate of 4% over the next 12 months, the matrix B which we expect would give the information for the month of August 2007 is

$$\begin{pmatrix} 1.04 \times 50 & 1.04 \times 12.4 & 1.04 \times 8 & 1.04 \times 6.1 \\ 1.04 \times 52 & 1.04 \times 13 & 1.04 \times 8.5 & 1.04 \times 6.3 \\ 1.04 \times 49.3 & 1.04 \times 12.5 & 1.04 \times 7.9 & 1.04 \times 6 \end{pmatrix} = \begin{pmatrix} 52 & 12.9 & 8.32 & 6.34 \\ 54.08 & 13.52 & 8.84 & 6.55 \\ 51.27 & 13 & 8.22 & 6.24 \end{pmatrix}$$

Matrices of this type, but decidedly much larger and filled with more data, are fundamental in studying the changes in prices of commonly used items. They are thus often used by government offices which gather and monitor such statistics. We have thus arrived, in a very natural way, at *the definition of the product of a number and a matrix of a given type as that matrix, of the same type, which has as its entries the product of the number with the corresponding entry in the original matrix.*

If we want to express the rule in a formal way, we say that, given a matrix $A = (a_{ij})$ and a number α, then

$$\alpha A = (\alpha\, a_{ij})$$

We have thus described the **product of a matrix by a number** (or by a **scalar**).

2.2 Row by column product

In the last chapter we saw an important use of row by column products of matrices. Now we would like to deepen that discussion with the help of some interesting examples.

Example 2.2.1. Mountain villages
Let's consider the following situation. Suppose we were in a mountain village and we wanted to get to another by means of certain mountain paths. Let's call, for brevity, the village we are starting from S, and the village where we would like to arrive A.

Looking at the map we notice that we will have to pass through one of four other villages, which we will call B_1, B_2, B_3, B_4. We also notice that from S to B_1 there are 3 possible routes, from S to B_2 there are 2 possible routes, from S to B_3 there are 4 possible routes, and from S to B_4 there is only one route. Moreover we notice that from B_1 to A there are 2 possible routes, from B_2 to A there are 5 possible routes, from B_3 to A there is only one possible route, and from B_4 to A there are 4 possible routes.

It is natural to ask the following question: how many different routes are there from S to A? The reasoning is not very difficult. It requires little reflection to realize that the total number of routes is the sum of the number of routes that pass through B_1 with the number of routes which pass through B_2 with the number of routes that pass through B_3 with the number of routes that pass through B_4.

And how many of these routes, for example, pass through B_1? We can go from S to B_1 with three possible routes and from B_1 to A with 2 possible routes and so it is clear that to go from S to A passing through B_1 we have $3 \times 2 = 6$ possible routes. We can repeat the same reasoning to count the possible routes through B_2, B_3 and B_4 and one concludes that the total number of possible routes is

$$3 \times 2 + 2 \times 5 + 4 \times 1 + 1 \times 4 = 24$$

Let's now try to find a mathematical model to describe what we have just described in words. First we can use a row matrix (or a vector) to encapsulate the information about the number of possible routes from S to B_1, B_2, B_3, and B_4 respectively. In this way we get a row matrix

$$M = (3 \quad 2 \quad 4 \quad 1)$$

Notice that the representation as a row matrix means that the unique row *represents the village S* while the four columns of the matrix *represent* the four villages B_1, B_2, B_3, B_4. In other words, with this type of representation we have implicitly decided that the rows (in this case there is only one) represent the villages where we start and the columns the villages where we can arrive. To maintain some coherence then, the paths from the villages B to the village A will thus be represented by a matrix of type $(4, 1)$, that is by a column matrix, more precisely by the following matrix

$$N = \begin{pmatrix} 2 \\ 5 \\ 1 \\ 4 \end{pmatrix}$$

For example, in matrix M the entry in position $(1, 3)$ represents the number of possible paths between S and B_3, in matrix N the entry in position $(2, 1)$ represents the number of possible routes between B_2 and A.

You are probably beginning to understand that in this case the mathematical model is the following. We define

$$M \cdot N = \begin{pmatrix} 3 & 2 & 4 & 1 \end{pmatrix} \begin{pmatrix} 2 \\ 5 \\ 1 \\ 4 \end{pmatrix} = 3 \times 2 + 2 \times 5 + 4 \times 1 + 1 \times 4 = 24$$

and then the total number of possible paths which connect S with A is the single entry in the matrix $M \cdot N$. In other words, forming the product $M \cdot N$ as suggested, we obtain a matrix of type $(1, 1)$, that is with one single entry, and that entry is precisely the number 24.

The example we just described admits many generalizations. In particular, the most obvious is that which one gets, for example, by considering 3 possible starting villages S_1, S_2, S_3 and two possible goals, villages A_1 and A_2. The number of possible paths between the villages S_3 and A_2, for example, one obtains by adding to the number of possible paths through B_1 the number of possible paths through B_2 plus... In other words, we repeat the reasoning before for each of the possible villages from which we can start with each of the possible villages where we can arrive.

Following the conventions we used before it's clear that the number of possible paths between the starting villages S_1, S_2, S_3 and the villages B_1, B_2, B_3, B_4 can be described by a matrix M of type $(3, 4)$ and that the number of possible paths between the villages B_1, B_2, B_3, B_4 and the villages A_1, A_2 are described by a matrix N of type $(4, 2)$.

If we do the calculations as we did in the preceding example, for each pair, (S_1, A_1), (S_1, A_2), (S_2, A_1), (S_2, A_2), (S_3, A_1), (S_3, A_2) we have a number. But then it seems natural to write those six numbers in a matrix of type $(3, 2)$, that, given the way it was constructed, it would be correct to call a row by column product between M and N. *We will write such a matrix using the symbol $M \cdot N$ or simply MN.*

The matrix MN, the row by column product of M and N, is thus of type $(3, 2)$ and the entry in position i, j represents the total number of the possible paths from the village S_i to the village A_j. Thus, if one has

$$M = \begin{pmatrix} 1 & 2 & 1 & 5 \\ 3 & 2 & 4 & 1 \\ 3 & 1 & 4 & 1 \end{pmatrix} \qquad N = \begin{pmatrix} 7 & 1 \\ 1 & 5 \\ 1 & 2 \\ 2 & 3 \end{pmatrix}$$

we obtain

$$MN = \begin{pmatrix} 20 & 18 \\ \mathbf{29} & 14 \\ 27 & 14 \end{pmatrix}$$

For example, the total number of possible paths between S_2 and A_1 is

$$3 \times 7 + 2 \times 1 + 4 \times 1 + 1 \times 2 = 29$$

The more adventurous reader might be asking how much more one can abstract and generalize the preceding reasoning. This is typical mathematical curiosity. But note, I don't mean this is an unusual happening. Many developments in mathematics begin with questions of this type, questions which appear to be without practical content. But that is only their appearance. . .

Let's think about this some more. Notice that this type of product was useful in the preceding chapter (see Section 1.4) in order to describe a system of linear equations using the formalism $A\mathbf{x} = \mathbf{b}$, where $A\mathbf{x}$ is precisely the row by column product of the coefficient matrix A and the column matrix $\mathbf{x}$.

Let's also observe that an essential condition which allows us to carry out the row by column product of two matrices A and B is that the number of the columns of A is equal to the number of rows of B. That is so because of the following: the number of columns of a matrix is the number of entries in each of its rows while the number of rows of a matrix is the number of the entries of each of its columns. The mathematical formalism of what was suggested by the preceding considerations is the following

Suppose $A = (a_{ij}) \in \text{Mat}_{r,c}(\mathbb{Q})$, $B = (b_{ij}) \in \text{Mat}_{c,d}(\mathbb{Q})$. We construct the matrix $A \cdot B = (p_{ij}) \in \text{Mat}_{r,d}(\mathbb{Q})$, by defining

$$p_{ij} = a_{i1}b_{1j} + a_{i2}b_{2j} + \cdots + a_{ic}b_{cj}$$

The matrix constructed in this way has r rows (like A) and d columns (like B) and is called the **row by column product** *of A and B. Often, for convenience, we write AB instead of $A \cdot B$.*

The fact that the entries of the two matrices are rational isn't relevant. What's important is that the entries are in the same numerical entity and, moreover, in that entity we can form both sums and products. For example, the same procedure would work perfectly if, instead of rational entries, we had real entries. Let's see some other examples.

Example 2.2.2. Let's try to find the matrix product of the following two matrices

$$A = \begin{pmatrix} 3 & 2 & 0 \\ 1 & 2 & 1 \\ 0 & 0 & -1 \\ 3 & 2 & 7 \\ 1 & 1 & 1 \\ 2 & 2 & 0 \end{pmatrix} \qquad B = \begin{pmatrix} 0 & -1 \\ 1 & 1 \\ 1 & 1 \end{pmatrix}$$

Notice that the number of columns of A coincides with the number of rows of B and is 3. So, we can go on and we get

$$A \cdot B = \begin{pmatrix} 3 \cdot 0 + 2 \cdot 1 + 0 \cdot 1 & 3 \cdot (-1) + 2 \cdot 1 + 0 \cdot 1 \\ 1 \cdot 0 + 2 \cdot 1 + 1 \cdot 1 & 1 \cdot (-1) + 2 \cdot 1 + 1 \cdot 1 \\ 0 \cdot 0 + 0 \cdot 1 + (-1) \cdot 1 & 0 \cdot (-1) + 0 \cdot 1 + (-1) \cdot 1 \\ 3 \cdot 0 + 2 \cdot 1 + 7 \cdot 0 & 3 \cdot (-1) + 2 \cdot 1 + 7 \cdot 1 \\ 1 \cdot 0 + 1 \cdot 1 + 1 \cdot 0 & 1 \cdot (-1) + 1 \cdot 1 + 1 \cdot 1 \\ 2 \cdot 0 + 2 \cdot 1 + 0 \cdot 0 & 2 \cdot (-1) + 2 \cdot 1 + 0 \cdot 1 \end{pmatrix} = \begin{pmatrix} 2 & -1 \\ 3 & 2 \\ -1 & -1 \\ 2 & 6 \\ 1 & 1 \\ 2 & 0 \end{pmatrix}$$

Notice that we have, as was expected from the general discussion, the number of rows in the matrix product equal to 6 (as was true for A) and the number of its columns is 2 (as was true for B).

Example 2.2.3. The weighted graph
Let's consider the following figure

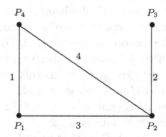

If, for the moment, we ignore the numbers in the figure, the part that remains is the placing of the 4 points P_1, P_2, P_3, P_4, and the fact that there are line segments which join some of them. For example P_1 is connected to P_2 and to P_4, while P_3 is connected to only P_2.

One understands immediately that a figure of this type could be a model for the description of many different things. It could, for example, represent the road connections between four towns P_1, P_2, P_3, P_4, and thus we see the possibility to generalize Example 2.2.1. It could also represent the number of electrical connections between four sub-stations P_1, P_2, P_3, P_4, and so on. Because of the importance and the generality of the concept, figures like these (without the numbers) are called **graphs** and have been studied intensively by mathematicians.

The numbers in the figure can represent, for example, the number of direct connections (in this case the graphs are called **weighted**). Thus, between P_4 and P_3 there are no direct connections while between P_1 and P_2 there are three.

Let's consider the problem of trying to collect all the numerical information which appears in the figure. One reasonable way to do this is certainly the following:

	P_1	P_2	P_3	P_4
P_1	0	3	0	1
P_2	3	0	2	4
P_3	0	2	0	0
P_4	1	4	0	0

Even better, we can get rid of the letters and only write the matrix

$$A = \begin{pmatrix} 0 & 3 & 0 & 1 \\ 3 & 0 & 2 & 4 \\ 0 & 2 & 0 & 0 \\ 1 & 4 & 0 & 0 \end{pmatrix}$$

It's useful to make some observations about the matrix A. First of all we have decided to declare 0 as the number of direct connections between any point and itself. We are describing, with this, a choice rather than a rule, a choice that, among other things, has its practical advantages, as we will shortly see. The creation of mathematical objects to describe phenomena is subject, like all human creations, to taste, mood and convention. For example, it's clear that one could use the number 1 instead of the 0 on the diagonal (see the Example 2.5.1) and that would have given us another interpretation, namely that every point has one direct connection with itself.

We find ourselves in a situation similar to that of the mathematician who decides that $2^0 = 1$. A priori, 2^0 should mean $2 \times 2 \cdots \times 2$ as many times as is indicated by the exponent, i.e. 0. But, that wouldn't mean anything and thus there is a certain liberty in how to define this. On the other hand, that liberty is quickly limited by the desire to extend, to this particular case, a well known property of exponents. One would like to have, for example, that $\frac{2^3}{2^3} = 2^{3-3} = 2^0$ and inasmuch as the left hand side is equal to 1, we get the convention that says that $2^0 = 1$.

Let's return to our matrix. We notice right away that we have a square matrix and, as noted, all the elements on the **principal diagonal** are zero. Notice that $a_{14} = a_{41} = 1$, and that $a_{23} = a_{32}$. In general we have the equality $a_{ij} = a_{ji}$ for any i, j. We could say that the matrix is "reflected" with respect to its principal diagonal. Such matrices are called **symmetric**. The formal definition is the following.

Let $A = (a_{ij})$ be a matrix.

(a) The **transpose** of A, denoted by A^{tr}, is the matrix that, for all i, j we have a_{ji} as the entry in position (i, j).

(b) The matrix $A = (a_{ij})$ is called symmetric if $a_{ij} = a_{ji}$ for all i, j. In other words, A is symmetric if $A = A^{\text{tr}}$.

Notice that the definition of symmetric matrices forces them to be square. Special examples of symmetric matrices are: the square matrices in which all the entries are zero; the identity matrix I_r (which we will see in a moment). Another important class of examples of symmetric matrices are the following.

Let $A = (a_{ij})$ be a square matrix. If $a_{ij} = 0$ whenever $i \neq j$, then we say that A is a **diagonal** matrix.

Let's return to our graph. Why is the matrix A that we have associated to the graph symmetric? The reason is that we have not oriented the connections, there is no *unique direction* which we have given, for example, to the 4 direct connections between P_2 and P_3. Thus it makes the same sense to say that there are also 4 direct connections between P_3 and P_2.

Since A is a square matrix of size 4×4, we can perform the row by column product of A with A and obtain a matrix which we (correctly) call A^2 and which is again a matrix of size 4×4.

$$A^2 = A \cdot A = \begin{pmatrix} 10 & 4 & 6 & 12 \\ 4 & 29 & 0 & 3 \\ 6 & 0 & 4 & 8 \\ 12 & 3 & 8 & 17 \end{pmatrix}$$

And now comes the interesting question! What can we deduce from the entries of A^2? The first thing that shouldn't surprise us is that A^2 is symmetric, given that we had a row by column product and A was symmetric. Moreover, let's try (for example) to interpret the entry, 12, in the $(1,4)$ position. Notice that we get it as

$$12 = 0 \times 1 + 3 \times 4 + 0 \times 0 + 1 \times 0$$

The interpretation is similar to that given to the routes that connected the mountain villages and generalizes that example. Let's see how. There are 0 direct connections between P_1 and P_1 (recall our convention) and 1 direct connection between P_1 and P_4. There are 3 direct connections between P_1 and P_2 and 4 direct connections between P_2 and P_4. There are 0 direct connections between P_1 and P_3 and 0 direct connections between P_3 and P_4. Finally, there is 1 direct connection between P_1 and P_4 and 0 direct connections between P_4 and P_4. Thus, there are exactly 0×1 connections of length 2 between P_1 and P_4 that pass through P_1, there are 3×4 that pass through P_2 and so on. In conclusion, the matrix A^2, that is the matrix obtained by performing the row by column product of A with itself, represents the number of connections of length 2 between any two points of the graph.

Example 2.2.4. The graph
Let's now consider the following graph

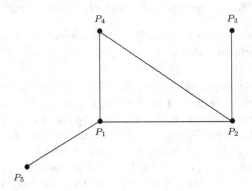

Notice that we have not written any numbers near the edges. We do this to express the fact there is either a direct connection, for example between P_5 and P_1, or that there is no connection, as for example between P_1 and P_3. This graph can represent the connections between a collection of machines. For example, it might be that P_1 represents a computer and that P_2, P_4 and P_5 three peripheral devices (say P_2 represents a keyboard, P_4 a printer and P_5 a monitor) while P_3 represents a peripheral device of the keyboard, for example a mouse, and that there is a direct connection between the keyboard (P_2) and the printer (P_4).

Using the same reasoning as in example 2.2.3 we can write the connection matrix A

$$A = \begin{pmatrix} 0 & 1 & 0 & 1 & 1 \\ 1 & 0 & 1 & 1 & 0 \\ 0 & 1 & 0 & 0 & 0 \\ 1 & 1 & 0 & 0 & 0 \\ 1 & 0 & 0 & 0 & 0 \end{pmatrix}$$

Here also we can take the square of A and obtain

$$A^2 = \begin{pmatrix} 3 & 1 & 1 & 1 & 0 \\ 1 & 3 & 0 & 1 & 1 \\ 1 & 0 & 1 & 1 & 0 \\ 1 & 1 & 1 & 2 & 1 \\ 0 & 1 & 0 & 1 & 1 \end{pmatrix}$$

It should not be too difficult for the reader to interpret the significance of A^2. For example, the fact that a_{45} is 1 signifies that there is exactly one connection of length 2 between the printer P_4 and the monitor P_5. In fact, examining the graph we can immediately exhibit the unique connection of length 2 which passes through the computer P_1.

2.3 How much does it cost to multiply two matrices?

Now that we have seen important uses of the matrix product, let's make a digression in order to discuss a fundamentally important question in *effective calculation*. We want to evaluate the price we have to pay, i.e. the computational cost, to produce the row by column product of two matrices. What does this mean?

Clearly it doesn't make sense to ask ourselves how much time it takes for a calculator to make a specific computation since the answer depends too much on the nature of the calculator. It would be like asking how long it would take for a car to make the trip between Montreal and Kingston on Highway 401.

However, is there something in the calculation of the product that doesn't depend on the calculator we use? In keeping with the metaphor we just introduced, it's clear that there is a piece of intrinsic data in the problem of driving along Highway 401 between Montreal and Kingston, and that is the *number of kilometers*. Thus, in calculating the product of matrices it will be useful to count *the number of elementary operations* that have to be performed.

Notice that keeping track only of the number of operations doesn't take into account the *unit cost* of each single operation. However, I don't want to enter into this extremely delicate discussion, even though it is at the very center of the area called **complexity theory**, a subject of great importance in modern Information Theory.

With this proviso, let's concentrate on counting the number of operations necessary to calculate a product. So, suppose we are given two matrices A and B and that A is of type (r, c) and B is of type (c, d). Their product is the matrix $A \cdot B$, of type (r, d), as seen in Section 2.2. Given that we have to perform rd row by column products in order to fill in all of the entries of $A \cdot B$, it will be enough to multiply rd by the number of operations it will cost us to make a single row by column product. The entry in position (i, j) of $A \cdot B$ is

$$a_{i1}b_{1j} + a_{i2}b_{2j} + \cdots + a_{ic}b_{cj}$$

The operations we have to perform are thus c multiplications and $c - 1$ sums. This number of operations has to be made for each entry of $A \cdot B$ and thus has to be made, as we said, rd times. In conclusion, the number of elementary operations to perform is

$$rdc \text{ products} \qquad \text{and} \qquad rd(c - 1) \text{ sums}$$

In particular, if the matrices A and B are square of type n, the number of operations to perform is

$$n^3 \text{ products} \qquad \text{and} \qquad n^2(n - 1) \text{ sums}$$

It's interesting to take a moment to think about what it means to say we have to perform n^3 products. Let's remember that ℓ^3 is the volume of a cube with side ℓ. Remember that this expression of the third degree in ℓ (or as we usually say, cubic in ℓ) has the following effect: if we double the side of the cube we increase the volume by a factor of eight. Analogously, for a square matrix A of type 5 the number of multiplications needed to calculate A^2 is $5^3 = 125$, while for a matrix of type 10, i.e. of twice the size, the number of multiplications to calculate its square is $10^3 = 1000$, i.e. eight times 125.

2.4 Some properties of the product of matrices

Since we are now familiar with the product of matrices, it's a good time to discover the fact that many of the properties that *everyone* knows about multiplication of numbers *don't hold* for the multiplication of matrices! In this section we will only give some numerical examples. The idea (hope) is that these examples will be enough to convince the reader that we are moving around in unfamiliar territory.

Let's start by considering two matrices A and B, A of type (r, c) and B of type (r', c'). We have already seen that in order to make the product $A \cdot B$ it's necessary that we have $c = r'$. But, this doesn't mean that we can also perform the product $B \cdot A$. In fact, to make that product we need another condition, more precisely we need that $c' = r$. In other words, in order to be able to perform the two multiplications $A \cdot B$ and $B \cdot A$ we need that A be of type (r, c) and that B be of type (c, r).

OK, let's suppose that A is of type (r, c) and that B is of type (c, r) and let's also suppose that $r \neq c$. In this case the product $A \cdot B$ is a matrix of type (r, r), i.e. a square matrix of type r, while $B \cdot A$ is a matrix of type (c, c), i.e. a square matrix of type c. Since $r \neq c$ there is no way that $A \cdot B = B \cdot A$. Let's look at an example. Let

$$A = \begin{pmatrix} 2 & 2 & 1 \\ 1 & 0 & 0 \end{pmatrix} \qquad B = \begin{pmatrix} 1 & -1 \\ -2 & 0 \\ 3 & 3 \end{pmatrix}.$$

Thus one has

$$A \cdot B = \begin{pmatrix} 1 & 1 \\ 1 & -1 \end{pmatrix} \qquad B \cdot A = \begin{pmatrix} 1 & 2 & 1 \\ -4 & -4 & -2 \\ 9 & 6 & 3 \end{pmatrix}$$

Are there cases in which both the products can be made and they both give matrices of the same type? From the discussion above it's clear that this possibility exists only when the two matrices are square and of the same type. But, even in this case we're in for a surprise. Consider the following example. Let

$$A = \begin{pmatrix} 2 & 2 \\ 1 & 0 \end{pmatrix} \qquad B = \begin{pmatrix} 1 & -1 \\ -2 & 0 \end{pmatrix}$$

Then one has

$$A \cdot B = \begin{pmatrix} -2 & -2 \\ 1 & -1 \end{pmatrix} \qquad B \cdot A = \begin{pmatrix} 1 & 2 \\ -4 & -4 \end{pmatrix}$$

Clearly we see that $A \cdot B \neq B \cdot A$. We can finish with the following affirmation.

The row by column product of matrices is not commutative.

It's a good idea to explore the row by column product a bit more, since the entire discussion up to this point has shown that we are dealing with an operation of fundamental importance.

Everyone knows that if a and b are two numbers different from 0 then their product ab is different from 0. This property is not maintained for matrices. In fact, it can happen that a power of a non-zero matrix is the matrix of all zeroes. A matrix with this last property is called **nilpotent**. For example, the matrix $A = \begin{pmatrix} 0 & 1 \\ 0 & 0 \end{pmatrix}$ is not the zero matrix, but $A^2 = A \cdot A = \begin{pmatrix} 0 & 0 \\ 0 & 0 \end{pmatrix}$ is.

Let's go forward with our investigations even if, at this moment, it's not clear why we are gathering this information. I can, however, assure you that the knowledge we are accumulating now will be very useful... and soon.

Everyone knows that the number 1 has the property that it is neutral with respect to the product of numbers. By this I mean that $1 \cdot a = a = a \cdot 1$ no matter what the number a is. Given this, the following question arises spontaneously to a mathematician: is there a matrix which behaves with respect to matrix products like the number 1 behaves with respect to number products? This could seem a useless question, but we will see right away that it is not. Let's consider the matrices

$$M = \begin{pmatrix} 1 & 2 & 1 & 5 \\ 3 & 2 & 4 & 1 \\ 3 & 0 & 4 & 1 \end{pmatrix} \quad \text{and} \quad N = \begin{pmatrix} 7 & 1 \\ 1 & 0 \\ 1 & 2 \\ 2 & 3 \end{pmatrix}$$

which we have already seen in the problem of the Mountain Villages. Moreover, let's consider the matrices

$$I_2 = \begin{pmatrix} 1 & 0 \\ 0 & 1 \end{pmatrix} \qquad I_3 = \begin{pmatrix} 1 & 0 & 0 \\ 0 & 1 & 0 \\ 0 & 0 & 1 \end{pmatrix} \qquad I_4 = \begin{pmatrix} 1 & 0 & 0 & 0 \\ 0 & 1 & 0 & 0 \\ 0 & 0 & 1 & 0 \\ 0 & 0 & 0 & 1 \end{pmatrix}$$

Doing the calculations one easily sees that

$$I_3 \cdot M = M = M \cdot I_4$$

and that

$$I_4 \cdot N = N = N \cdot I_2$$

It thus seems that there are many matrices which act like the number 1 and we also see that we have to make a clear distinction between multiplication on the left and multiplication on the right.

Clearly we are dealing with very special matrices. If A is a matrix of type (r, s) and I_r and I_s are the matrices of type (r, r) and (s, s) having all entries on the diagonal equal to 1 and all other entries equal to zero, then we can form the product $I_r \cdot A$ and the result is A and if we form the product $A \cdot I_s$, the result is also A. This property of the matrices I_r and I_s, is similar to that of the number 1 which leaves a product unchanged, encourages us to give these matrices a name. We will call them **identity matrices**, respectively, of type r and s. If there is no danger of ambiguity we will indicate them simply using the letter I.

We end this section with some good news! We know that in the integers the following formulas hold

$$a + (b + c) = (a + b) + c \qquad (ab)c = a(bc) \qquad a(b + c) = ab + bc$$

Said in another way, **addition of integers is associative, multiplication of integers is associative**, and **multiplication of integers distributes over addition of integers**.

It turns out that these formulas are also valid for matrices when we consider the sum and the row by column product of matrices as the operations indicated in the formulas (always supposing that the operations can be carried out with matrices). For example, the reader can verify that if

$$A = \begin{pmatrix} 1 & 2 & 3 \\ 0 & -1 & 3 \end{pmatrix} \qquad B = \begin{pmatrix} 0 & 1 & 1 \\ 1 & -1 & 3 \end{pmatrix} \qquad C = \begin{pmatrix} 1 & 2 & 3 \\ 0 & -1 & 3 \end{pmatrix}$$

then one has

$$A + (B + C) = \begin{pmatrix} 1 & 2 & 3 \\ 0 & -1 & 3 \end{pmatrix} + \begin{pmatrix} 1 & 3 & 4 \\ 1 & -2 & 6 \end{pmatrix} = \begin{pmatrix} 2 & 5 & 7 \\ 1 & -3 & 9 \end{pmatrix}$$

$$(A + B) + C = \begin{pmatrix} 1 & 3 & 4 \\ 1 & -2 & 6 \end{pmatrix} + \begin{pmatrix} 1 & 2 & 3 \\ 0 & -1 & 3 \end{pmatrix} = \begin{pmatrix} 2 & 5 & 7 \\ 1 & -3 & 9 \end{pmatrix}$$

The reader can verify, for example, that if

$$A = \begin{pmatrix} 1 & 2 & 3 \\ 0 & -1 & 3 \end{pmatrix} \qquad B = \begin{pmatrix} 2 \\ 3 \\ -10 \end{pmatrix} \qquad C = \begin{pmatrix} 7 \\ 0 \\ -11 \end{pmatrix}$$

then one has

$$A(B + C) = \begin{pmatrix} 1 & 2 & 3 \\ 0 & -1 & 3 \end{pmatrix} \begin{pmatrix} 9 \\ 3 \\ -21 \end{pmatrix} = \begin{pmatrix} -48 \\ -66 \end{pmatrix}$$

$$AB + AC = \begin{pmatrix} -22 \\ -33 \end{pmatrix} + \begin{pmatrix} -26 \\ -33 \end{pmatrix} = \begin{pmatrix} -48 \\ -66 \end{pmatrix}$$

The reader can verify, for example, that if

$$A = \begin{pmatrix} 1 & 2 & 3 \\ 0 & -1 & 3 \end{pmatrix} \qquad B = \begin{pmatrix} 2 \\ 3 \\ -10 \end{pmatrix} \qquad C = (-1 \quad -2)$$

then one has

$$A(BC) = \begin{pmatrix} 1 & 2 & 3 \\ 0 & -1 & 3 \end{pmatrix} \begin{pmatrix} -2 & -4 \\ -3 & -6 \\ 10 & 20 \end{pmatrix} = \begin{pmatrix} 22 & 44 \\ 33 & 66 \end{pmatrix}$$

and also

$$(AB)C = \begin{pmatrix} -22 \\ -33 \end{pmatrix} (-1 \quad -2) = \begin{pmatrix} 22 & 44 \\ 33 & 66 \end{pmatrix}$$

Actually, I'm not sure if the reader is ready to consider these facts as *good news*. What's the positive aspect of these properties? Without getting into the nitty gritty of these delicate mathematical questions it will be enough to reflect on the fact that having these properties permits us to work with great *liberty* when making calculations. Perhaps the reader is still not convinced that these properties represent *good news*? If the gentle reader has the patience to continue I believe that what follows will be convincing.

2.5 Inverse of a matrix

Mathematicians love to use (what they call) **number fields** such as $\mathbb{Q}$ (the field of rational numbers), $\mathbb{R}$ (the field of real numbers), $\mathbb{C}$ (the field of complex numbers). Why? Largely because of an important property that they have in common, namely that **every non-zero element has an inverse** under multiplication. That is a property that the set of integers $\mathbb{Z}$ doesn't have (algebraists would like to correct me and speak of the *ring of integers*) because, for example, the number 2 doesn't have an integer as inverse.

Inasmuch as we previously introduced an operation of product of two matrices and we have also seen that there exist identity matrices, it is natural to ask ourselves if there exist matrix inverses. Probably there are some readers who would advance a doubt that such a question is, in fact, natural. Other readers will say that these are the curious kind of questions that are typical of mathematicians who seem to enjoy studying abstract structures.

That's not quite right. Inverse matrices play a role of fundamental importance even in applications and in an attempt to convince the reader of the validity of this assertion we will see an interesting example which will permit us to make a side trip into the *fascinating* land of algebra (beware, when a mathematician uses the word fascinating there could be some danger on the horizon...).

By making a simple calculation, it's easy to see that if A is a matrix and there is another matrix B for which $AB = I_r = BA$, then A has to be a square matrix of type r. A matrix like B will be denoted, conveniently, by A^{-1}, and it is not difficult to prove that A^{-1} (if it exists) is unique. Moreover, mathematicians know how to prove that if A is a square matrix and B is a left inverse, i.e. $BA = I$, then it is also a right inverse, i.e. $AB = I$. Analogously, if B is a right inverse, then it is also a left inverse. The simplest example of a matrix which has an inverse is I_r. In fact, $I_r I_r = I_r$ and hence $I_r^{-1} = I_r$, in other words, I_r *is its own inverse*.

Let's take a break from this formal mathematical language and *reflect* a bit on what we have just seen (the reason we've used italics to write the word "reflect" I'll leave to the specialists). There are actions which, done once have a certain effect but done twice erase the effect. Some examples are: taking a playing card and turning it over; flipping a switch to turn on a light or even taking the negative of a number. In fact, if you turn over the card twice it ends up in the same position as at the start and if you flip the light switch twice and you started with the light on it will still be on or if you started with the light off it will still be off. Finally, the negative of the negative of a number is the number itself.

Is there a mathematical way to capture the essence of this reflection? Let's try. Suppose we had a mathematical world made of only two symbols, which no one would prohibit us from calling 0 and 1; moreover we could establish the convention that said that 0 corresponded to doing nothing and 1 to doing something! The preceding discussion could then conveniently be interpreted by saying that

$$1 + 1 = 0$$

It seems a bit strange, but instead the discussion becomes even more interesting when one realizes that one can go on and discover that the following equalities also make some sense

$$1 + 0 = 0 + 1 = 1 \quad \text{and} \quad 0 + 0 = 0$$

interpreting 0 and 1 as above. In fact, the equality $1 + 0 = 1$ can be interpreted as *doing something and then doing nothing is the same as doing something*, something which we could agree makes sense.

What we have just said can be synthesized in the following *addition table* on the set of the two symbols 0 and 1

+	0	1
0	0	1
1	1	**0**

Only the zero in the right bottom position seems strange, inasmuch as we would expect to find the number 2 in that place. And what if we were to put alongside this table of sums the *usual* multiplication table?

×	0	1
0	0	0
1	0	1

What kind of significance is it possible to give to this product operation? As for the set $\{0,1\}$ to which we have given these operations of sum and product; in what way does it resemble, for example, the set $\mathbb{Q}$, the field of rational numbers?

As far as the first question is concerned, we will shortly give a concrete reason for this strange structure. The second question sounds a bit odder! But, one should observe that, just like for the rational numbers, every number different from 0 has an inverse; in fact the unique element different from 0, i.e. 1, is its own inverse and we see that from the fact that $1 \times 1 = 1$. Thus, there is an analogy between these objects. Mathematicians give a name to this structure which is made up of two numbers and two operations, they call it $\mathbb{Z}_2$ (or $\mathbb{F}_2$) and they observe that one is dealing with a number field just like $\mathbb{Q}$. Both the names $\mathbb{Z}_2$ and $\mathbb{F}_2$ contain the symbol 2, which indicates that in this set $1 + 1 = 0$, which is a bit like saying that $2 = 0$. As you can see, we have returned to the point where we began, i.e. to the fact that there are situations (even practical situations) in which doing something twice is like not doing anything at all, i.e. in which $2 = 0$!

Now we are going to make yet another leap ahead. Let's consider a square matrix A of type r whose entries are in $\mathbb{Z}_2$ and look for, if it exists, the inverse A^{-1}. But, what is the point of doing this? We'll soon see an example which will make sense of this strange idea.

Example 2.5.1. Let's turn on the lights
Let's suppose that we have an electric circuit represented by the following graph

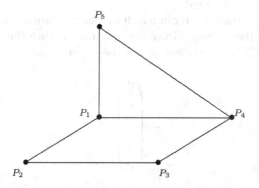

The vertices P_1, P_2, P_3, P_4, P_5 of the graph represent devices consisting of a lamp and a switch; the edges of the graph represent direct connections obtained, for example, by electric wires. Each switch can either be ON or

OFF and each lamp can be either ON or OFF. If we either turn on or turn off a switch that action changes the situation of the lamp where that switch is located as well as the lamps *adjacent* to that switch, i.e. those corresponding to vertices directly connected to that switch.

Let's see what happens with an example. Suppose that lamps P_1 and P_3 are on and that lamps P_2, P_4 and P_5 are off and suppose we turn the switches (once) at P_3 and P_4. What happens? What is the state of the 5 lamps? Let's look at the lamp at P_1. Turning the switch at P_3 doesn't alter the situation of that lamp since P_1 and P_3 are not adjacent, on the other hand turning the switch at P_4 does change the situation of the lamp at P_1, since P_4 and P_1 are adjacent. Thus, the state of the lamp at P_1 is changed by turning the two switches and since the lamp at P_1 was on at the beginning, it will be off after. We can do the same sort of analysis for each lamp and the final result will be: the lamp at P_1 is off, at P_2 on, at P_3 on, at P_4 off and at P_5 on.

If the idea is now sufficiently clear, the next step is to find a good mathematical model for this kind of situation. We begin with the observation that we can place all the information about the connections in the graph as entries in a symmetric matrix of type 5, just as we did in Section 2.2. We obtain the matrix

$$A = \begin{pmatrix} 1 & 1 & 0 & 1 & 1 \\ 1 & 1 & 1 & 0 & 0 \\ 0 & 1 & 1 & 1 & 0 \\ 1 & 0 & 1 & 1 & 1 \\ 1 & 0 & 0 & 1 & 1 \end{pmatrix}$$

Recall that, for example, the entry 1 in position $(4, 3)$ means that there is a direct connection between P_4 and P_3, while the entry 0 in position $(3, 5)$ means that there is no direct connection between P_3 and P_5. Also observe that, unlike Example 2.2.3, the nature of this example suggests that we put 1's all along the main diagonal.

Now consider a column matrix in which we insert information about which switches we would like to turn. Given that we want to turn the switches only at P_3 and P_4, it will be convenient to use the column matrix

$$V = \begin{pmatrix} 0 \\ 0 \\ 1 \\ 1 \\ 0 \end{pmatrix}$$

Having arrived at this point, one has a beautiful application of the matrix product. We can interpret both A and V as matrices having entries in $\mathbb{Z}_2$ and we can perform the product because A is of type 5 and V is of type $(5, 1)$.

One obtains

$$
A \cdot V = \begin{pmatrix} 1 & 1 & 0 & 1 & 1 \\ 1 & 1 & 1 & 0 & 0 \\ 0 & 1 & 1 & 1 & 0 \\ 1 & 0 & 1 & 1 & 1 \\ 1 & 0 & 0 & 1 & 1 \end{pmatrix} \begin{pmatrix} 0 \\ 0 \\ 1 \\ 1 \\ 0 \end{pmatrix} = \begin{pmatrix} 1 \\ 1 \\ 0 \\ 0 \\ 1 \end{pmatrix}
$$

Let's see why we get the result indicated above and also try to understand the significance of the product $A \cdot V$. Let's figure out the entry in position $(3, 1)$ of the product. We get that entry by taking the product of the third row of A with the column V. One thus has $0 \times 0 + 1 \times 0 + 1 \times 1 + 1 \times 1 + 0 \times 0 = 0$ (recall the addition table, in $\mathbb{Z}_2$ one has $1 + 1 = 0$).

Now comes the really interesting part. What does the sum of the products $0 \times 0 + 1 \times 0 + 1 \times 1 + 1 \times 1 + 0 \times 0$ really mean? Let's remember that the third row concerned the place P_3. The first summand 0×0 can be read in the following way: the switch at P_3 *is not* directly connected with the position P_1 (hence the reason for the zero in position $(3, 1)$ of the matrix A) which interacts with the fact that the status of the switch at P_1 *was not* changed (hence the reason for the 0 in the first place of the matrix V). The result of this interaction is clearly that the state of the lamp at P_3 was *not* changed.

The second summand 1×0 can be read in the following way: the switch at P_3 *is* directly connected with the place P_2 (hence the reason for the 1 in position $(3, 2)$ of the matrix A) and this interacts with the fact that the status of the switch at P_2 was *not* altered (hence the reason for the zero in position $(2, 1)$ of the matrix V). The result of this interaction is that the state of the lamp in P_3 is *not* altered.

The third summand 1×1 can be interpreted in the following way: the switch at P_3 *is* directly connected with the place P_3 (hence the reason for the 1 in position $(3, 3)$ of the matrix A) and that interacts with the fact that the status of the switch at P_3 *is* altered (hence the reason for the 1 in position $(3, 1)$ of the matrix V). The result of this interaction is that the state of the lamp at P_3 *is* altered.

Now it should be clear how to go on and explain all the five summands. We have to perform the sum of the five actions in order to see the total effect on the lamp at P_3. We have $0 + 0 + 1 + 1 + 0$ and hence the state of the lamp at P_3 was altered twice. The result of the sum is 0, in other words, the state of the lamp at P_3 is unchanged.

One can interpret all the other products in an analogous way and conclude that this product of matrices with entries in $\mathbb{Z}_2$ is a good mathematical model for the problem of seeing how the electrical system is altered by changing the status of one or more of the switches.

It's also very interesting to take a look at the multiplication table of $\mathbb{Z}_2$, i.e. at the following table

$$
\begin{array}{c|cc}
\times & 0 & 1 \\
\hline
0 & 0 & 0 \\
1 & 0 & 1
\end{array}
$$

especially now that it has taken on a very concrete meaning. In fact, $0 \times 1 = 0$ puts together two things from the mathematical model: first that there is no direct connection between the first and the second position and combines that with the fact that changing the status of the switch in the second position doesn't alter the state of the lamp in the first position, and so on.

One should observe the subtlety that in this model 1×0 has a significance that is completely different from 0×1, this despite the fact that both of these give the result 0, which saves the commutativity of the product. If you enjoy this last observation that is a clear symptom that you are suffering from *mathematical madness*.

Let's go one step further. Suppose we want to figure out what we have to do with the switches in order to get some desired result, given that we know the current state of the lamps. The difference between the final state and the initial state gives us a table of the changes we want to make in the states. For example if, at the beginning, all the lamps were off and we wanted, in the end, to have all the lamps on, that means we want to change the state of all of the lamps. Thus, we want to make a series of turns of the switches in order to obtain the column matrix

$$
\mathbf{b} = \begin{pmatrix} 1 \\ 1 \\ 1 \\ 1 \\ 1 \end{pmatrix}
$$

What we don't know, i.e. the unknowns, in this problem is what we have to do to each of the switches. I.e. we have five unknowns x_1, x_2, x_3, x_4, x_5. When all is said and done we see that what we want are the solutions to the system of linear equations

$$
A\mathbf{x} = \mathbf{b} \quad \text{i.e.} \quad
\begin{pmatrix}
1 & 1 & 0 & 1 & 1 \\
1 & 1 & 1 & 0 & 0 \\
0 & 1 & 1 & 1 & 0 \\
1 & 0 & 1 & 1 & 1 \\
1 & 0 & 0 & 1 & 1
\end{pmatrix}
\begin{pmatrix} x_1 \\ x_2 \\ x_3 \\ x_4 \\ x_5 \end{pmatrix}
=
\begin{pmatrix} 1 \\ 1 \\ 1 \\ 1 \\ 1 \end{pmatrix}
$$

Once again we find ourselves facing a system of linear equations S having five equations and five unknowns. Moreover, the notion of an inverse comes into play in an essential manner. In fact, if there is a matrix inverse for A, then we have that

$$
\mathbf{x} = A^{-1}A\mathbf{x} = A^{-1}\mathbf{b}
$$

and thus we have found the solution! In our case we don't know yet how to calculate the inverse, but we can check that

$$A^{-1} = \begin{pmatrix} 1 & 1 & 0 & 1 & 0 \\ 1 & 0 & 0 & 0 & 1 \\ 0 & 0 & 0 & 1 & 1 \\ 1 & 0 & 1 & 1 & 0 \\ 0 & 1 & 1 & 0 & 1 \end{pmatrix}$$

thus

$$\mathbf{x} = A^{-1}\mathbf{b} = \begin{pmatrix} 1 \\ 0 \\ 0 \\ 1 \\ 1 \end{pmatrix}$$

is the solution. In other words, if the lamps are all off then in order to turn them all on we have to change the status of the switches in positions P_1, P_4 and P_5, as you can actually verify easily with the help of the graph.

We'll end this important section with a question. Why did we look for the inverse of the matrix in this example and not try to solve the system of linear equations directly? And, if there had not been an inverse for A, how would we have found a solution? It's probably a good idea to get used to the fact that science, like life, has more questions than answers. But, in this case we can consider ourselves fortunate, because in the next chapter we will get answers to these two questions.

Exercises

Exercise 1. Calculate, where possible, the product of the following pairs of matrices

(a) $A = \begin{pmatrix} 0 & 1 \\ 0 & 0 \end{pmatrix}$ $A' = \begin{pmatrix} 0 & 2 \\ 0 & 0 \end{pmatrix}$

(b) $A = \begin{pmatrix} 0 & 1 & 0 \\ 0 & 0 & 1 \end{pmatrix}$ $A' = \begin{pmatrix} 0 & 2 & 2 \\ 0 & 0 & 4 \end{pmatrix}$

(c) $A = \begin{pmatrix} 0 & 1 & 0 \\ 0 & 0 & 1 \end{pmatrix}$ $A' = \begin{pmatrix} 0 & 2 \\ 0.3 & 0 \\ 0.2 & 5 \end{pmatrix}$

Exercise 2. Does it cost more to multiply two matrices A, B which are square of type 6 or two matrices A, B of type $(4, 5)$ and $(5, 11)$ respectively?

Exercise 3. Is it true that one can multiply a matrix by itself only if it is a square matrix?

Exercise 4. Let I be the identity matrix of type 2.

(a) Find all the solutions of the matrix equation $X^2 - I = 0$, i.e. all the square matrices $A \in \mathrm{Mat}_2(\mathbb{R})$ for which $A^2 - I = 0$.
(b) Is it true that there are an infinite number of solutions?
(c) Is it true that in all the solutions we must have the equality $|a_{11}| = |a_{22}|$?

Exercise 5. Let $A \in \mathrm{Mat}_n(\mathbb{R})$ and let I be the identity matrix of type n.

(a) Show that if $A^3 = 0$ then $I + A$ and $I - A + A^2$ are inverses of each other.
(b) Is it true that if there is a natural number k such that $A^k = 0$ then $I + A$ is invertible?
(c) Is it always true that a matrix of type $I + A$ is invertible?

Exercise 6. Calculate A^3 in each of the following cases

$$A = \begin{pmatrix} 0 & 1 \\ 0 & 0 \end{pmatrix} \qquad A = \begin{pmatrix} 1 & 0 \\ 0 & 1 \end{pmatrix}$$

Exercise 7. Let $A = \begin{pmatrix} 1 & -1 \\ 1 & 0 \end{pmatrix}$. Verify that $A^6 = I$.

Exercise 8. Consider the matrices

$$A = \begin{pmatrix} 1 & 1 & 1 \\ 0 & -1 & 3 \\ 2 & -1 & 1 \end{pmatrix} \qquad B = \begin{pmatrix} 1 & a & b \\ a & -1 & b \\ b & -b & a \end{pmatrix}$$

Find, if they exist, real numbers a, b such that the two matrices A, B commute (i.e. such that $A \cdot B = B \cdot A$).

Exercise 9. Let A be a matrix of type (r, c) and B a matrix of type (c, s).

(a) Is it true that if A has a row of zeroes then AB also has a row of zeroes?

(b) Is it true that if A has a column of zeroes then AB also has a column of zeroes?

Exercise 10. Construct a simple example, similar to that of Example 2.5.1 in which, starting from a particular configuration of switches and lamps (some lamps on and the rest off) in which it is not possible to arrive at the state in which all the lamps are on.

Exercise 11. Let A be a matrix.

(a) Is it true that $(A^{tr})^{tr} = A$, no matter what A is?

(b) Is it true that if A^{tr} is symmetric then A is also symmetric?

Exercise 12. Let A be a matrix with real entries.

(a) Prove that the elements of the diagonal of $A^{tr}A$ are all non-negative.

(b) If $(A^{tr})^{tr} = I$, can one describe A^{-1} without having to calculate it?

Exercise 13. Let's consider matrices with rational entries and suppose that there is the same cost for each operation between numbers (addition or multiplication).

(a) Calculate the computational cost of forming the product of two square diagonal matrices of type n.

(b) Calculate the computational cost of finding A^2 when A is a symmetric matrix of type n.

Exercise 14. Consider the diagonal matrices in $\mathrm{Mat}_2(\mathbb{Q})$, i.e. the matrices $A_{a,b} = \begin{pmatrix} a & 0 \\ 0 & b \end{pmatrix}$ con $a, b \in \mathbb{Q}$.

(a) Prove that for every $a \in \mathbb{Q}$ the matrix $A_{a,a}$ commutes with all the matrices in $\mathrm{Mat}_2(\mathbb{Q})$.

(b) Is the same statement true for the matrix $A_{1,2}$?

(c) Prove that any diagonal matrix commutes with all the diagonal matrices.

The following exercises are a bit different from the ones above. In fact the word **Exercise** *is preceded by the symbol* **ⓠ**. *This means that in order to work the exercise one needs a calculator (see the Introduction and the Appendix) or, in any case, the use of a calculator is strongly advised. In the last chapter we will see a more sophisticated method to solve Exercises 15 and 16 which uses the idea of eigenvalue.*

ⓠ Exercise 15. Calculate A^{100} for the following matrices

(a) $A = \begin{pmatrix} 1 & 1 \\ 0 & 2 \end{pmatrix}$

(b) $A = \begin{pmatrix} 3 & 0 \\ 2 & -1 \end{pmatrix}$

ⓠ Exercise 16. Consider the following matrices

$$A = \begin{pmatrix} 0 & 1 & -\frac{1}{2} \\ 0 & 0 & 12 \\ 3 & \frac{1}{5} & 8 \end{pmatrix} \qquad B = \begin{pmatrix} \frac{1}{3} & 1 & 1 \\ 2 & 1 & -21 \\ 0 & \frac{3}{4} & 1 \end{pmatrix}$$

and prove the following equalities

$$A^{13} = \begin{pmatrix} \dfrac{281457596383971}{6250} & \dfrac{8243291212479289}{1000000} & \dfrac{2579611252226942479}{2000000} \\ \dfrac{1883521814429871}{3125} & \dfrac{13791079790208861}{125000} & \dfrac{431570585554290003}{250000} \\ \dfrac{431570585554290003}{1000000} & \dfrac{394993103775412801}{5000000} & \dfrac{154508738617589077}{125000} \end{pmatrix}$$

$$B^{13} = \begin{pmatrix} \dfrac{2075574373808189}{3265173504} & \dfrac{-2771483961974593}{272097792} & \dfrac{-34285516978000235}{2176782336} \\ \dfrac{-22589583602079623}{1088391168} & \dfrac{-7482652061373805}{725594112} & \dfrac{155899288381048673}{725594112} \\ \dfrac{46412434031431}{120932352} & \dfrac{-2468698236647575}{322486272} & \dfrac{-872661281513917}{80621568} \end{pmatrix}$$

calculators are not intelligent,
but they think they are

3

Solutions of Systems of Linear Equations

> *in theory there is no difference*
> *between theory and practice,*
> *in practice there is*

In this chapter we come to grips with the question of how to solve, *in practice*, systems of linear equations. Our *strategy* will be to gather a certain number of observations which will allow us to form a *strategy*. Given that many mathematicians often use the adjective *clear* for something which usually is far from clear but is tiresome *to prove* we will *prove* our good intention to avoid this bad habit by beginning with an observation that is totally *clear*.

Recall that we used the symbol I to indicate the identity matrix: our first observation is that a system of linear equations of the form $I\mathbf{x} = \mathbf{b}$ clearly has the solution $\mathbf{x} = \mathbf{b}$. In fact since we have the equality $I\mathbf{x} = \mathbf{x}$ then $\mathbf{x} = \mathbf{b}$ is clear. Thus, we have *solved this system of linear equations*.

But, it's hard to imagine ever having the good fortune to find oneself with such a simple system of linear equations. In general we don't even expect the matrix of coefficients to be a square matrix. So, how do we proceed? The basic idea is to replace the original system of linear equations with another system of linear equations which has the same solutions but whose coefficient matrix is *more similar* to the identity matrix and thus is the coefficient matrix of a system of linear equations which is *easier* to solve.

This idea brings us to the study of *elementary matrices* (which will reinforce the importance of the row by column product) and allows us to describe an algorithm called the Gauss Method based on the choice of special elements called *pivots*. We'll be able to calculate the inverse of a matrix, when it exists. We'll also make a digression on the computational cost of the Gauss Method. In addition we will learn when and how to decompose a square matrix as a product LU, where L and U are two special triangular matrices. Finally we will witness the arrival on stage of certain numbers called *determinants*. These are certain *non linear* expressions which play an essential role in linear algebra.

Robbiano L.: Linear Algebra for everyone
© Springer-Verlag Italia 2011

Wait a minute. Didn't we say that this was to be a chapter of practical methods? OK, let's quit the chatting and get down to work.

3.1 Elementary Matrices

From now on we will use the convention of calling two systems of linear equations **equivalent** if they have exactly the same solutions. The word equivalent was not chosen at random. Mathematicians love this word, and for good reasons. Without going into detail, just let me say that the notion of an **equivalence relation** is at the base of a large part of the formal structure of mathematics. Inasmuch as we are interested in solving a system of linear equations, having this notion of equivalence gives us the freedom to substitute one system for another one that is equivalent to it.

The reader is likely to ask: what do I gain from doing this? I will begin my discussion by responding to this question.

To get started let me say that we plan to do a certain number of elementary operations which will transform our original system of linear equations into an equivalent system which is *much easier to solve*. Let's begin with some simple observations.

(a) If we interchange two equations of a system of linear equations we get an equivalent system of linear equations.

(b) If we multiply an equation in a given system of linear equations by a number different from zero, we get an equivalent system of linear equations.

(c) If we replace an equation in a linear system by an equation which is the sum of the original equation and a multiple of another equation of the system, we get an equivalent system of linear equations.

The three operations mentioned above are called **elementary operations** on the given system of linear equations.

Let's look, in detail, at a couple of examples.

Example 3.1.1. A system having two equations and two unknowns
Consider the following system of linear equations

$$\begin{cases} 2x_1 - 3x_2 = 2 \\ x_1 + x_2 = 4 \end{cases} \qquad (1)$$

The solutions to this system of linear equations are the same as the solutions to the system of linear equations

$$\begin{cases} x_1 + x_2 = 4 \\ 2x_1 - 3x_2 = 2 \end{cases} \qquad (2)$$

which is obtained by interchanging the two equations in (1), and are also the same as the solutions to

$$\begin{cases} x_1 + x_2 = 4 \\ -5x_2 = -6 \end{cases} \quad (3)$$

which is obtained by substituting, for the second equation in (2), the second equation minus two times the first equation, and are also the same as the solutions to

$$\begin{cases} x_1 + x_2 = 4 \\ x_2 = \frac{6}{5} \end{cases} \quad (4)$$

which is obtained by multiplying the second equation in (3) by the number $-\frac{1}{5}$, and are also the same as the solutions to

$$\begin{cases} x_1 = \frac{14}{5} \\ x_2 = \frac{6}{5} \end{cases} \quad (5)$$

which is obtained by substituting for the first equation in (4), the first equation minus the second equation. In this case, given that the system (5) is of the form $I\mathbf{x} = \mathbf{b}$, it can be solved immediately and we have done what we set out to do.

Example 3.1.2. A system having two equations and three unknowns
Consider the following system of linear equations

$$\begin{cases} 2x_1 - 3x_2 + x_3 = 2 \\ x_1 + x_2 - 5x_3 = 4 \end{cases} \quad (1)$$

Its solutions are the same as that of the system of linear equations

$$\begin{cases} x_1 + x_2 - 5x_3 = 4 \\ 2x_1 - 3x_2 + x_3 = 2 \end{cases} \quad (2)$$

which is obtained by interchanging the two equations of (1), and moreover has the same solutions as the system of linear equations

$$\begin{cases} x_1 + x_2 - 5x_3 = 4 \\ -5x_2 + 11x_3 = -6 \end{cases} \quad (3)$$

which is obtained by replacing the second equation in (2) by the second equation minus twice the first, and this system has the same solutions as

$$\begin{cases} x_1 + x_2 - 5x_3 = 4 \\ x_2 - \frac{11}{5}x_3 = \frac{6}{5} \end{cases} \quad (4)$$

which is obtained by multiplying the second equation in (3) by $-\frac{1}{5}$, and moreover has the same solutions as

$$\begin{cases} x_1 & - \frac{14}{5}x_3 = \frac{14}{5} \\ x_2 - \frac{11}{5}x_3 = \frac{6}{5} \end{cases} \tag{5}$$

which is obtained by replacing the first equation in (4) with the first equation minus the second.

In both the preceding examples we obtained, for each change of equations, a new system which was equivalent to the preceding one and hence for which the solutions to (1) were the same as the solutions to (5). We also saw that the system of linear equations (5) in Example 3.1.1 was of the form $I\mathbf{x} = \mathbf{b}$ and thus we immediately saw the solution. The system of linear equations (5) of Example 3.1.2 is quite different and we will speak about it in Section 3.6

We'll return, later on, to the actual solutions of the systems of equations but, for now, let's analyze the various steps we did in the calculations above. One notices immediately that the modifications we made on the systems of linear equations *only involved the matrices which are part of the system* and certainly not the names of the unknowns.

It will be convenient to give a name to the transformations of the matrices which correspond to the elementary transformations of the system of linear equations.

We will call them **elementary transformations** *of a matrix. To sum up they are:*

(a) *The interchange of two rows.*
(b) *The multiplication of a row by a number different from zero.*
(c) *Replacing a row by the row you get adding the given row to a multiple of another row.*

Thus, the elementary operations on a system of linear equations $A\mathbf{x} = \mathbf{b}$ can be done by doing elementary operations on the matrices A and $\mathbf{b}$. To see how that works, let's revisit Example 3.1.1 and follow each step of that example.

The passage from (1) to (2) was done by interchanging the rows of A and $\mathbf{b}$ and hence gives us the matrices

$$A_2 = \begin{pmatrix} 1 & 1 \\ 2 & -3 \end{pmatrix} \qquad \mathbf{b}_2 = \begin{pmatrix} 4 \\ 2 \end{pmatrix}$$

which are, respectively, the matrices of coefficients and the constant term of the system of linear equations (2). Analogously the passage from (2) to (3) gives us the matrices

$$A_3 = \begin{pmatrix} 1 & 1 \\ 0 & -5 \end{pmatrix} \qquad \mathbf{b}_3 = \begin{pmatrix} 4 \\ -6 \end{pmatrix}$$

which are, respectively, the matrices of coefficients and the constant term of the system of linear equations (3). The passage from (3) to (4) give us the matrices

$$A_4 = \begin{pmatrix} 1 & 1 \\ 0 & 1 \end{pmatrix} \qquad \mathbf{b}_4 = \begin{pmatrix} 4 \\ \frac{6}{5} \end{pmatrix}$$

which are, respectively, the coefficient matrix and constant term of the system of linear equations (4). The passage from (4) to (5) gives us the matrices

$$A_5 = \begin{pmatrix} 1 & 0 \\ 0 & 1 \end{pmatrix} \qquad \mathbf{b}_5 = \begin{pmatrix} \frac{14}{5} \\ \frac{6}{5} \end{pmatrix}$$

which are, respectively, the coefficient matrix and constant term of the system of linear equations (5), which give us the explicit solution to the system of equations.

We know that mathematics is rich in surprises and abundant in marvels and now it offers us one that is quite interesting. Let's look again at Example 3.1.1 and consider the identity matrix I of type 2. If we interchange the two rows of I we get the matrix $\begin{pmatrix} 0 & 1 \\ 1 & 0 \end{pmatrix}$ which we will call E_1. Now, let's multiply A and $\mathbf{b}$ on the left by this matrix. We get

$$E_1 A = \begin{pmatrix} 1 & 1 \\ 2 & -3 \end{pmatrix} \qquad E_1 \mathbf{b} = \begin{pmatrix} 4 \\ 2 \end{pmatrix}$$

The *amazing thing* consists in the fact that these are, respectively, the coefficient matrix and constant term of the system of linear equations (2), which we can thus write as

$$E_1 A \, \mathbf{x} = E_1 \mathbf{b}$$

The rule of interchanging two rows is thus implemented by multiplying on the left by the matrix we get by doing the corresponding interchange on the rows of the identity matrix. There's more: analogous matrices can also be obtained for the other elementary operations on a system of linear equations. In conclusion, we have at hand the following set of rules which are *applicable to any matrix*, even if it is not square. Let A be a matrix with r rows.

(a) Denoting by E the matrix one gets by interchanging rows i and j of the matrix I_r, the matrix product EA is the matrix one obtains by interchanging rows i and j of A.

(b) Denoting by E the matrix one gets by multiplying the i-th row of I_r by the constant γ, the matrix product EA is the matrix one obtains by multiplying the i-th row of A by the constant γ.

(c) Denoting by E the matrix one gets by adding to the i-th row of I_r the j-th row of I_r multiplied by the constant γ, the matrix product EA is the matrix one obtains from A by adding to its i-th row, γ times its j-th row.

Given that we have called the above mentioned operations on the rows of the matrix A elementary operations, the matrices that we obtain by carrying out the above mentioned operations on the identity matrix will be called **elementary matrices**. There is an elegant way to sum up what was said above, namely:

Every elementary operation on the rows of a matrix A can be implemented by multiplying A on the left by the corresponding elementary matrix.

Having recognized the importance of the elementary matrices, we should examine them in some detail. Let's consider the example of the elementary matrix obtained by interchanging the second and fourth row of the identity matrix I_4, i.e.

$$E = \begin{pmatrix} 1 & 0 & 0 & 0 \\ 0 & 0 & 0 & 1 \\ 0 & 0 & 1 & 0 \\ 0 & 1 & 0 & 0 \end{pmatrix}$$

If we perform the multiplication $EE = E^2$ we obtain the identity matrix I_4. It's easy to understand why since multiplying E on the left by E has the effect of interchanging the second and fourth row of E and thus bringing us back to the matrix we started with, namely the identity matrix I_4.

Consider the example of the elementary matrix obtained by multiplying the second row of the identity matrix I_4 by the number 2. We get

$$E = \begin{pmatrix} 1 & 0 & 0 & 0 \\ 0 & 2 & 0 & 0 \\ 0 & 0 & 1 & 0 \\ 0 & 0 & 0 & 1 \end{pmatrix}$$

and let's consider the analogous operation in which the multiplication is by $\frac{1}{2}$

$$E' = \begin{pmatrix} 1 & 0 & 0 & 0 \\ 0 & \frac{1}{2} & 0 & 0 \\ 0 & 0 & 1 & 0 \\ 0 & 0 & 0 & 1 \end{pmatrix}$$

If we perform the multiplication $E'E$ it's clear we will get the identity matrix I_4. Now let's consider the elementary matrix obtained by adding to the second row of the identity matrix I_4, 3 times the third row. The result is

$$E = \begin{pmatrix} 1 & 0 & 0 & 0 \\ 0 & 1 & 3 & 0 \\ 0 & 0 & 1 & 0 \\ 0 & 0 & 0 & 1 \end{pmatrix}$$

and if we consider the analogous elementary operation, in which the multiplication is done by -3, we get the elementary matrix

$$E' = \begin{pmatrix} 1 & 0 & 0 & 0 \\ 0 & 1 & -3 & 0 \\ 0 & 0 & 1 & 0 \\ 0 & 0 & 0 & 1 \end{pmatrix}$$

If we perform the multiplication $E'E$ we obtain the identity matrix I_4. Should we be surprised? Certainly not, especially if we realize that adding to the second row of the identity matrix the third row multiplied by -3 has the effect of undoing the operation we made on the identity matrix of adding to the second row 3 times the third row. In view of these examples, and the discussion around then, the following fact should come as no surprise.

(1) **Elementary matrices are invertible and have as inverses elementary matrices.**

(2) **If $Ax = b$ is a system of r linear equations and the matrices $E_1, E_2, \ldots, E_{m-1}, E_m$ are elementary matrices of type r, then the system of linear equations**

$$E_m E_{m-1} \cdots E_2 E_1 Ax = E_m E_{m-1} \cdots E_2 E_1 b$$

is equivalent to the system of linear equations $Ax = b$.

Let's go over the various steps we performed in Example 3.1.1, the example with which we began this section. The system of linear equations is

$$Ax = b \qquad \text{where} \qquad A = \begin{pmatrix} 2 & -3 \\ 1 & 1 \end{pmatrix} \qquad b = \begin{pmatrix} 2 \\ 4 \end{pmatrix}$$

The passage from (1) to (2) is obtained by multiplying the matrices A and b on the left by the matrix $E_1 = \begin{pmatrix} 0 & 1 \\ 1 & 0 \end{pmatrix}$. We end up with the matrices

$$A_2 = \begin{pmatrix} 1 & 1 \\ 2 & -3 \end{pmatrix} = E_1 A \qquad b_2 = \begin{pmatrix} 4 \\ 2 \end{pmatrix} = E_1 b$$

Analogously, the passage from (2) to (3) is obtained by multiplying the matrices A_2 and b_2, always on the left, by the matrix $E_2 = \begin{pmatrix} 1 & 0 \\ -2 & 1 \end{pmatrix}$. This time we end up with the matrices

$$A_3 = \begin{pmatrix} 1 & 1 \\ 0 & -5 \end{pmatrix} = E_2 A_2 = E_2 E_1 A \qquad b_3 = \begin{pmatrix} 4 \\ -6 \end{pmatrix} = E_2 b_2 = E_2 E_1 b$$

The passage from (3) to (4) is obtained by multiplying the matrices A_3 and b_3 on the left by the matrix $E_3 = \begin{pmatrix} 1 & 0 \\ 0 & -\frac{1}{5} \end{pmatrix}$. We now end up with the matrices

$$A_4 = \begin{pmatrix} 1 & 1 \\ 0 & 1 \end{pmatrix} = E_3 A_3 = E_3 E_2 E_1 A \qquad b_4 = \begin{pmatrix} 4 \\ \frac{6}{5} \end{pmatrix} = E_3 b_3 = E_3 E_2 E_1 b$$

The passage from (4) to (5) is obtained by multiplying the matrices A_4 and b_4 on the left by the matrix $E_4 = \begin{pmatrix} 1 & -1 \\ 0 & 1 \end{pmatrix}$. This time we end up with the matrices

$$A_5 = \begin{pmatrix} 1 & 0 \\ 0 & 1 \end{pmatrix} = E_4 A_4 = E_4 E_3 E_2 E_1 A \qquad b_5 = \begin{pmatrix} \frac{14}{5} \\ \frac{6}{5} \end{pmatrix} = E_4 b_4 = E_4 E_3 E_2 E_1 b$$

The net effect of all of this is that the elementary operations on matrices can be interpreted as products (on the left) by elementary matrices. In our example we obtained $A_5 = I$ and hence we have an equivalent system with an explicit solution. But, we note that with this way of doing things we have gotten more than we might have suspected; in particular, we obtain $I = A_5 = E_4 E_3 E_2 E_1 A$ and get an explicit description of the inverse of A as a product of elementary matrices, namely

$$A^{-1} = E_4 E_3 E_2 E_1$$

If we carry out the computations we get

$$A^{-1} = \begin{pmatrix} \frac{1}{5} & \frac{3}{5} \\ -\frac{1}{5} & \frac{2}{5} \end{pmatrix} = \frac{1}{5} \begin{pmatrix} 1 & 3 \\ -1 & 2 \end{pmatrix}$$

The usefulness of the formalism of the product matrix interpretation of elementary row operations is that it has allowed us not only to find the solutions of a system of linear equations, but at the same time to calculate the inverse of the coefficient matrix of that system. This observation, we shall see, is of great importance since the coefficient matrix of the system of linear equations (1) is also the coefficient matrix of *any system of linear equations of the type* $A\mathbf{x} = \mathbf{b}'$, *where we can vary* $\mathbf{b}'$ *as we wish*.

Such a system of linear equations has a unique solution, namely $A^{-1}\mathbf{b}'$. Thus, solving one system of linear equations we have, at the same time, solved many others. This is another manifestation of the power of mathematics. It's worth emphasizing the fact that this kind of result is possible because we introduced the mathematical notation $A\mathbf{x} = \mathbf{b}$ and hence used the row by column product.

3.2 Square Linear Systems, Gaussian Elimination

We begin with the general problem of solving a system of linear equations $A\mathbf{x} = \mathbf{b}$, in the case where A is a square matrix. Such systems of linear equations will be called **square systems of linear equations**.

When the coefficient matrix is invertible we have, in a certain sense, already solved the problem. We have said that with the hypothesis that the coefficient matrix is invertible there is a unique solution, namely $\mathbf{x} = A^{-1}\mathbf{b}$. But, here we have to be careful. Given a square system of linear equations, we don't know ahead of time if the coefficient matrix A is invertible or not. We will discover that only *after and not before* we have solved the system of linear equations. When we discover that A is invertible then certainly the solution will be $A^{-1}\mathbf{b}$, but the fundamental point is that we don't, in general, calculate the inverse of A but rather calculate the solution $A^{-1}\mathbf{b}$. This last phrase does not contradict what was said at the end of the last section. We emphasized the fact that if we calculate A^{-1}, we can easily calculate the solutions to all the systems of linear equations $A\mathbf{x} = \mathbf{b}$, varying $\mathbf{b}$ as we wish. But the question is very different if we want to calculate only the solutions to one system of linear equations. Let's return for a moment to Example 3.1.1. Recall that after some transformations one arrives at the equivalent system

$$\begin{cases} x_1 + x_2 = 4 \\ \quad\ x_2 = \frac{6}{5} \end{cases} \qquad (4)$$

At this point the coefficient matrix is not the identity matrix but the system can be easily solved. In fact the second equation gives us that $x_2 = \frac{6}{5}$. *Substituting* that into the first equation we get $x_1 = 4 - \frac{6}{5} = \frac{14}{5}$.

This observation suggest the following method, which has come to be known as the **Gauss Method** or the **Method of Gaussian Reduction**, for solving the system of linear equations $A\mathbf{x} = \mathbf{b}$.

(a) Using elementary operations on the matrix A, one obtains a matrix A' which is **upper triangular**, i.e. a matrix with the property that all the elements below the main diagonal are zero.

(b) If A' has all the elements on the main diagonal non-zero, then the equivalent system of linear equations $A'\mathbf{x} = \mathbf{b}$ can be solved by **back substitution**.

Example 3.2.1. Let's examine, in detail, the system of linear equations that grew out of Example 2.5.1. We want to solve the following system

$$\begin{pmatrix} 1 & 1 & 0 & 1 & 1 \\ 1 & 1 & 1 & 0 & 0 \\ 0 & 1 & 1 & 1 & 0 \\ 1 & 0 & 1 & 1 & 1 \\ 1 & 0 & 0 & 1 & 1 \end{pmatrix} \begin{pmatrix} x_1 \\ x_2 \\ x_3 \\ x_4 \\ x_5 \end{pmatrix} = \begin{pmatrix} 1 \\ 1 \\ 1 \\ 1 \\ 1 \end{pmatrix}$$

but remember that we are working in $\mathbb{Z}_2$ and thus that $1 + 1 = 0$ and $-1 = 1$. Inasmuch as $a_{11} = 1$, we can change things so that all the numbers under a_{11}, in the first column, are equal to zero. In fact, it's enough to perform the following three elementary operations: add to the second row the first row, add to the fourth row the first row, add to the fifth row the first row. One obtains the equivalent system

$$\begin{pmatrix} 1 & 1 & 0 & 1 & 1 \\ 0 & 0 & 1 & 1 & 1 \\ 0 & 1 & 1 & 1 & 0 \\ 0 & 1 & 1 & 0 & 0 \\ 0 & 1 & 0 & 0 & 0 \end{pmatrix} \begin{pmatrix} x_1 \\ x_2 \\ x_3 \\ x_4 \\ x_5 \end{pmatrix} = \begin{pmatrix} 1 \\ 0 \\ 1 \\ 0 \\ 0 \end{pmatrix}$$

Interchanging the second and the third rows we get

$$\begin{pmatrix} 1 & 1 & 0 & 1 & 1 \\ 0 & 1 & 1 & 1 & 0 \\ 0 & 0 & 1 & 1 & 1 \\ 0 & 1 & 1 & 0 & 0 \\ 0 & 1 & 0 & 0 & 0 \end{pmatrix} \begin{pmatrix} x_1 \\ x_2 \\ x_3 \\ x_4 \\ x_5 \end{pmatrix} = \begin{pmatrix} 1 \\ 1 \\ 0 \\ 0 \\ 0 \end{pmatrix}$$

and then we can make all the numbers under a_{22}, in the second column, equal to zero. In order to do that it's enough to perform the following elementary operations: add the second row to the fourth row and then add the second row to the fifth row. One obtains the following equivalent system of linear equations

$$\begin{pmatrix} 1 & 1 & 0 & 1 & 1 \\ 0 & 1 & 1 & 1 & 0 \\ 0 & 0 & 1 & 1 & 1 \\ 0 & 0 & 0 & 1 & 0 \\ 0 & 0 & 1 & 1 & 0 \end{pmatrix} \begin{pmatrix} x_1 \\ x_2 \\ x_3 \\ x_4 \\ x_5 \end{pmatrix} = \begin{pmatrix} 1 \\ 1 \\ 0 \\ 1 \\ 1 \end{pmatrix}$$

Now let's do the following elementary operation: add to the fifth row the third row. We get the following equivalent system

$$\begin{pmatrix} 1 & 1 & 0 & 1 & 1 \\ 0 & 1 & 1 & 1 & 0 \\ 0 & 0 & 1 & 1 & 1 \\ 0 & 0 & 0 & 1 & 0 \\ 0 & 0 & 0 & 0 & 1 \end{pmatrix} \begin{pmatrix} x_1 \\ x_2 \\ x_3 \\ x_4 \\ x_5 \end{pmatrix} = \begin{pmatrix} 1 \\ 1 \\ 0 \\ 1 \\ 1 \end{pmatrix}$$

Now doing back substitution one gets: $x_5 = 1$, $x_4 = 1$, $x_3 = 0 - x_4 - x_5 = 0$, $x_2 = 1 - x_3 - x_4 = 0$, $x_1 = 1 - x_2 - x_4 - x_5 = 1$ and thus, in conclusion, that the solution is $(1, 0, 0, 1, 1)$, which agrees with what we saw at the end of Example 2.5.1.

Let's take a moment to make a very important observation. In the preceding example, we could do the back substitution as soon as we obtained a coefficient matrix which was upper triangular with *non-zero elements on the main diagonal*. The reader should make a sincere effort to understand the importance of both of these facts, i.e. that the coefficient matrix has to be upper triangular and that the elements on the main diagonal have to not be zero.

Let's examine another example.

Example 3.2.2. Non-invertible square matrices
Let's consider the system of linear equations $A\mathbf{x} = \mathbf{b}$, where

$$A = \begin{pmatrix} 1 & 2 & -4 \\ 3 & 0 & 2 \\ 5 & 4 & -6 \end{pmatrix} \qquad \mathbf{b} = \begin{pmatrix} 0 \\ -1 \\ 1 \end{pmatrix}$$

Using the Gauss Method, we can replace the second row with the second row minus three times the first row, and we can replace the third row with the third row minus five times the first row. We get

$$A_2 = \begin{pmatrix} 1 & 2 & -4 \\ 0 & -6 & 14 \\ 0 & -6 & 14 \end{pmatrix} \qquad \mathbf{b}_2 = \begin{pmatrix} 0 \\ -1 \\ 1 \end{pmatrix}$$

If we now replace the third row with the third row minus the second row, we obtain

$$A_3 = \begin{pmatrix} 1 & 2 & -4 \\ 0 & -6 & 14 \\ 0 & 0 & 0 \end{pmatrix} \qquad b_3 = \begin{pmatrix} 0 \\ -1 \\ 2 \end{pmatrix}$$

At this point we see that the last equation has become $0 = 2$. We can thus conclude that the original system of linear equations doesn't have a solution.

Did the reader notice that at each reduction step in the last example, we were always using a non-zero entry on the main diagonal? When there wasn't one, we used the elementary operation of interchanging a pair of rows in order to get one. If we had been unable to do this, the procedure would have stopped. Thus, in Gaussian reduction, a nonzero entry on the main diagonal plays a central role, in basketball terms one would say that it plays a pivotal role. In fact, not surprisingly, such an entry is called a **pivot**.

We are now going to make a purely mathematical digression in order to clarify an important point. We want to see how and when the Gauss Method works and if it's true that we can ultimately decide (using that method) whether a matrix is invertible. So let A be a square matrix. An important observation is the following.

If a row of a matrix consists of only zeroes then the matrix is not invertible.

Mathematicians love to use **proofs by contradiction** in order to establish certain facts. We now see a proof by contradiction in action so that the reader can have an idea of how mathematics expands its area of control. In this case we want to be completely certain that if a row of a matrix consists entirely of zeroes then the matrix cannot be invertible or, equivalently, that if a matrix is invertible then none of its rows is the zero row (i.e. in each row there is at least one entry that is not zero). For those interested, here is the proof.

In order to prove the assertion one can reason in the following way. Suppose that the row of zeroes in the matrix A is the ith row. If there were an inverse B for A we would have $AB = I$ and so the ith row of AB would not consist entirely of zeroes. But, the ith row of AB is obtained by multiplying the ith row of A by the columns of B, and thus the ith row of AB is the zero row. We have arrived at a contradiction. In conclusion it is not possible for A to have a row of zeroes. With a completely analogous proof we can prove that if a column of A consists entirely of zeroes then A is not invertible.

Another important fact, whose easy proof we will give shortly (this time it will be a direct proof rather than a proof by contradiction), is the following.

The product of two invertible matrices A and B is an invertible matrix and one has $(AB)^{-1} = B^{-1}A^{-1}$.

To prove this we can proceed as follows.

Let A and B be the two given matrices. It will be enough to verify the equalities $B^{-1}A^{-1}AB = B^{-1}IB = B^{-1}B = I$ and that will finish the proof.

We can thus deduce that the matrix

$$A = \begin{pmatrix} 1 & 2 & -4 \\ 3 & 0 & 2 \\ 5 & 4 & -6 \end{pmatrix}$$

of Example 3.2.2 is not invertible, given that A_3 is not invertible and A_3 is the product of A by elementary matrices which are invertible. Notice also that the system of linear equations $Ax = b$ does not have any solutions, inasmuch as the equivalent system of linear equations $A_3x = b_3$ contains the equation $0 = 2$.

Can one conclude, then, that if A is not invertible then the systems of linear equations $Ax = b$ never have a solution? The answer is definitely not, i.e. we cannot draw such a conclusion from the fact that A is not invertible. In fact, we will see very soon that the answer depends on the value of b. Just think about the system of linear equations $Ax = b$ where

$$A = \begin{pmatrix} 1 & 1 \\ 1 & 1 \end{pmatrix} \qquad b = \begin{pmatrix} 2 \\ 2 \end{pmatrix}$$

which is equivalent to

$$A_2 = \begin{pmatrix} 1 & 1 \\ 0 & 0 \end{pmatrix} \qquad b = \begin{pmatrix} 2 \\ 0 \end{pmatrix}$$

and hence is equivalent to the single equation $x_1 + x_2 = 2$, which clearly has an infinite number of solutions. In this case the square system of linear equations which we started with is equivalent to a non-square system of linear equations. We will discuss non-square systems of linear equations in the next section.

We now make another important observation. In order to see directly if a square matrix A of type n is invertible, we can reason in the following way: our problem is to find a matrix X so that $AX = I_n$ and we can view this as an equation where A is known and X is unknown. By the definition of equality for matrices, the columns of the two matrices AX and I_n have to be the same. The j-th column of the matrix AX is Ax_j where x_j is the j-th column of the matrix X. So, the column Ax_j has to be equal to the j-th column of the identity matrix. All this amounts to is trying to find a solution to a system of linear equations whose coefficient matrix is A and whose column of constant terms is the j-th column of I_n.

So, the possibility of finding X is equivalent to the possibility of solving n systems of linear equations all having the same coefficient matrix A and each having, as its vector of constants, the various columns of the identity matrix I_n.

It's not hard to prove (but we won't do it here) that a square matrix A is invertible if and only if at every step of the Gaussian reduction one either finds a pivot which is not zero or we can make an opportune exchange of rows in order to obtain a non-zero pivot. Thus, with reduction and possible exchanges of rows, one arrives at an upper triangular matrix with all the elements on the main diagonal not zero.

On the other hand, Gaussian reduction and possible exchanges of rows bring us, in any case, to an upper triangular matrix. Putting together all of these facts we arrive at the following conclusion.

A square matrix is invertible if and only if the Gauss Method transforms it into an upper triangular matrix which has all of its elements on the main diagonal different from zero.

Thus we have revealed the *mystery* of the invertible matrix! When we start to solve a square system of linear equations, using the Gauss Method, we don't know if the coefficient matrix is invertible or not. But, we discover that along the way, precisely when we arrive at the triangular form. On the other hand, if we are only interested in solving the system of linear equations and we have arrived at a triangular form with all the elements on the main diagonal non-zero, then while we know that the coefficient matrix is invertible, we can solve the system by substitution without necessarily calculating the inverse.

3.3 Effective Calculation of Matrix Inverses

Having arrived at this point, it will be useful to stop and make some observations on the actual calculation of the inverse of a square matrix A, when A is invertible of type n. At the end of the Section 3.1 we said that the use of elementary matrices permits us to calculate the inverse of A and in Section 3.2 we observed that, in general, one doesn't actually calculate the inverse in order to solve a system of linear equations, but rather one uses the method of Gauss to do that. We have also observed that, at times, it can be useful to calculate the inverse of A, above all if we want to solve several systems of linear equations with the same coefficient matrix A. So, it is reasonable to stop and see how one can actually calculate A^{-1}. In a certain sense we have already seen how to do this at the end of Section 3.1. In fact we saw, with an example, that if we wrote down all the elementary operations $E_1, E_2, \ldots, E_r$ that transformed the matrix A into the identity matrix, then $A^{-1} = E_r E_{r-1} \cdots E_1$. But, in practice, *we don't have to multiply the elementary matrices.*

In fact, we saw at the end of Section 3.2 that the inverse of A may be thought of as the solution to the *matrix equation* $AX = I_n$, where X is an unknown matrix. We observe that from the relation $AX = I_n$ one deduces

$$E_r E_{r-1} \cdots E_1 A X = E_r E_{r-1} \cdots E_1 I_n \qquad (1)$$

If we suppose that $E_r E_{r-1} \cdots E_1 A = I_n$, then

$$X = E_r E_{r-1} \cdots E_1 I_n = E_r E_{r-1} \cdots E_1 \qquad (2)$$

Let's reread very carefully what is written in formulas (1) and (2). Note first that we wrote: if $E_r E_{r-1} \cdots E_1 A = I_n$ then, in other words, if the elementary operations described by the elementary matrices $E_1, E_2, \ldots, E_{r-1}, E_r$ bring A to the identity matrix, then the inverse of A (which is the matrix $E_r E_{r-1} \cdots E_1 I_n$) is the matrix that one obtains from the identity matrix by performing on it the *same elementary operations*. We have discovered the following rule.

If the elementary operations that we perform on the matrix A to transform it into I_n are performed on the identity matrix, then we get A^{-1}.

Let's look at an example in detail.

Example 3.3.1. Let's calculate the inverse
We consider the following square matrix of type 3,

$$A = \begin{pmatrix} 1 & 2 & 1 \\ 2 & -1 & 6 \\ 1 & 1 & 2 \end{pmatrix}$$

Let's put into practice what we said above and thus calculate the inverse of A, assuming that A is invertible, something that (for the moment) we don't know.

So, we will try to transform A into an upper triangular matrix with all diagonal entries equal to 1 using elementary operations and *at the same time use the same operations on the identity matrix.*
We use the entry in position $(1, 1)$ as a pivot and reduce to zero all the elements of the matrix *under* the pivot

$$\begin{pmatrix} 1 & 2 & 1 \\ 0 & -5 & 4 \\ 0 & -1 & 1 \end{pmatrix} \qquad \begin{pmatrix} 1 & 0 & 0 \\ -2 & 1 & 0 \\ -1 & 0 & 1 \end{pmatrix}$$

Now interchange the second and third rows

$$\begin{pmatrix} 1 & 2 & 1 \\ 0 & -1 & 1 \\ 0 & -5 & 4 \end{pmatrix} \qquad \begin{pmatrix} 1 & 0 & 0 \\ -1 & 0 & 1 \\ -2 & 1 & 0 \end{pmatrix}$$

Now let's use the position $(2, 2)$ as a pivot and reduce the element in the matrix *under* the pivot to zero

$$\begin{pmatrix} 1 & 2 & 1 \\ 0 & -1 & 1 \\ 0 & 0 & -1 \end{pmatrix} \qquad \begin{pmatrix} 1 & 0 & 0 \\ -1 & 0 & 1 \\ 3 & 1 & -5 \end{pmatrix}$$

Now we are going to do something that we don't need to do when we solve a system of linear equations. Using the technique of reduction, we will transform the triangular matrix into a diagonal matrix.

We use the entry in position $(3, 3)$ as a pivot and reduce all the elements of the matrix *above* the pivot to zero

$$\begin{pmatrix} 1 & 2 & 0 \\ 0 & -1 & 0 \\ 0 & 0 & -1 \end{pmatrix} \qquad \begin{pmatrix} 4 & 1 & -5 \\ 2 & 1 & -4 \\ 3 & 1 & -5 \end{pmatrix}$$

We use the entry in position $(2, 2)$ as a pivot and reduce the element *above* it to zero

$$\begin{pmatrix} 1 & 0 & 0 \\ 0 & -1 & 0 \\ 0 & 0 & -1 \end{pmatrix} \qquad \begin{pmatrix} 8 & 3 & -13 \\ 2 & 1 & -4 \\ 3 & 1 & -5 \end{pmatrix}$$

We multiply the second and third rows by -1

$$\begin{pmatrix} 1 & 0 & 0 \\ 0 & 1 & 0 \\ 0 & 0 & 1 \end{pmatrix} \qquad \begin{pmatrix} 8 & 3 & -13 \\ -2 & -1 & 4 \\ -3 & -1 & 5 \end{pmatrix}$$

At this point we see that the matrix A has been transformed into the identity matrix, while the identity matrix has been transformed into the matrix A^{-1}. We have the following equality

$$A^{-1} = \begin{pmatrix} 8 & 3 & -13 \\ -2 & -1 & 4 \\ -3 & -1 & 5 \end{pmatrix}$$

A particularly curious reader can easily verify the identity

$$AA^{-1} = A^{-1}A = I_3$$

and thus be completely convinced that we have, indeed, calculated the inverse of A.

Example 3.3.2. Let's calculate the inverse. . . if we can
Let's consider the following square matrix of type 3

$$A = \begin{pmatrix} 1 & 1 & 1 \\ 2 & 2 & 4 \\ 1 & 1 & 4 \end{pmatrix}$$

As in the preceding example, let's try to calculate the inverse of A, assuming that A is invertible (something we don't know yet).

We thus try to do the elementary operations that will transform A into an upper triangular matrix with all entries on the main diagonal equal to 1 and, at the same time, perform the same elementary operations on the identity matrix.

Let's get to work. We use the entry in position $(1,1)$ as a pivot and reduce to zero all the elements *under* it

$$\begin{pmatrix} 1 & 1 & 1 \\ 0 & 0 & 2 \\ 0 & 0 & 3 \end{pmatrix} \qquad \begin{pmatrix} 1 & 0 & 0 \\ -2 & 1 & 0 \\ -1 & 0 & 1 \end{pmatrix}$$

We use the entry in position $(2,3)$ as a pivot and reduce all the elements *under* it to zero

$$\begin{pmatrix} 1 & 1 & 1 \\ 0 & 0 & 2 \\ 0 & 0 & 0 \end{pmatrix} \qquad \begin{pmatrix} 1 & 0 & 0 \\ -2 & 1 & 0 \\ 2 & -\frac{3}{2} & 1 \end{pmatrix}$$

What's happened? We see that the third row of the matrix is zero and so the method of Gauss stops for lack of another pivot. Sadly we understand that the matrix we started with is not invertible. I said "sadly" because the operations we made on the identity matrix that brought it to the matrix on the right, have been useless. A waste of effort! But, was there some way of knowing *beforehand* that the matrix A was not invertible? In Section 3.7 we will discuss an answer to that question.

3.4 How much does Gaussian Elimination cost?

Let's make a *Genoese* digression in light of what we saw in Section 2.3 and try to figure out the cost of the Gauss Method, i.e. let's try to calculate how many elementary operations one must perform in order to find the solution to a square system of linear equations $A\mathbf{x} = \mathbf{b}$ when A is invertible of type n. In order to simplify things a bit we will assume that there is *no cost in exchanging rows*. We thus have to sum the costs of the following operations:

(1)   reduction to one of the first pivot and reduction to zero of the elements under the first pivot;

(2)   reduction to one of the second pivot and reduction to zero of the elements under the second pivot;

   . . .

(n-1) reduction to one of the $(n-1)$-st pivot and reduction to zero of the element under the $(n-1)$-st pivot;

(n)   reduction to one of the n-th pivot.

At this point the matrix is upper triangular with all elements on the main diagonal equal to 1 and all that is left to do is figure out the cost of the various back substitutions. More precisely, we have to evaluate how much the following operations cost:

(1)   substitute the value of x_n into the next to last equation and deduce
      the value of x_{n-1};
(2)   substitute into the third from last equation the values of x_n and x_{n-1}
      and deduce the value of x_{n-2};
      $\cdots$
(n-1) substitute into the first equation all the values we've found for x_n,
      $x_{n-1}, \ldots, x_2$ and deduce the value of x_1.

Let's first calculate the cost of the reduction to triangular form having all
elements on the main diagonal equal to 1.

(1)   Reduction to one of the first pivot and to zero of the elements under
      the first pivot.
      We have to divide each entry of the first row by the pivot. This accounts
      for n divisions. After this the pivot is now 1. For each row different from
      the first we add an appropriate multiple of the first row to it. There
      are $n - 1$ of these rows and for each we have to do n multiplications
      and n additions. In addition to these operations on A we have to add
      those on **b**. In other words we have to add 1 more division and $n - 1$
      more multiplications and $n - 1$ more additions.
(2)   Reduction to one of the second pivot and reduction to zero of the
      elements under the second pivot.
      Doing the same sort of reasoning as above one sees that we need to
      do $(n - 1)$ divisions, $(n - 1)(n - 2)$ multiplications and $(n - 1)(n - 2)$
      additions on A. On **b** we have to do 1 division, $n - 2$ multiplications
      and $n - 2$ additions.
      $\cdots$
(n-1) Reduction to zero of the element under the $(n - 1)$-st pivot.
      Again, reasoning as above, we see that we have to do 2 divisions, 2
      multiplications and 2 additions on A. On **b** we have to do 1 division,
      1 multiplication and 1 addition.
(n)   Reduction to one of the n-th pivot.
      We do 1 division on A and 1 division on **b**.

Adding it all up we find that the reduction to triangular form with all the
elements on the main diagonal equal to 1 requires:
   $n + (n - 1) + \cdots + 1$ divisions on A, and n divisions on **b**
   $n(n - 1) + (n - 1)(n - 2) + \cdots + 2$ multiplications on A, $(n - 1) + \cdots + 1$
multiplications on **b**
   $n(n-1)+(n-1)(n-2)+\cdots+2$ additions on A, $(n-1)+\cdots+1$ additions
on **b**.
One can show that
$$n + (n - 1) + \cdots + 1 = \frac{(n + 1)n}{2}$$
and that
$$n(n - 1) + (n - 1)(n - 2) + \cdots + 2 \cdot 1 = \frac{n^3 - n}{3}$$

Putting these various pieces together we can evaluate how much it costs to put the matrix in upper triangular form with all 1's on the main diagonal.

The cost is: $\frac{(n+1)n}{2}$ divisions, $\frac{n^3-n}{3}$ multiplications, and $\frac{n^3-n}{3}$ additions.

Inasmuch as the multiplications and the divisions are the biggest part of this sum, and for these the summand $\frac{n^3}{3}$ is the most relevant, one says: to bring a square matrix of type n to triangular form with 1's on the main diagonal has cost whose order of magnitude is $O(\frac{n^3}{3})$.

To these operations we have to add the operations on **b**, which are n divisions, $\frac{n(n-1)}{2}$ multiplications and $\frac{n(n-1)}{2}$ additions, whose total cost is $O(\frac{n^2}{2})$.

Now let's calculate the cost of the substitutions.

(1)   Substitution in the next to last equation to find the value of x_{n-1}.
One has to do one multiplication and 1 addition.
(2)   Substitution into the third from last equation to find the value of x_{n-2}.
One has to do two multiplications and two additions.

. . .

(n-1) Substitution into the first equation in order to deduce the value of x_1.
One has to do $n-1$ multiplications and $n-1$ additions.

In total the second part needs:
$n - 1 + (n - 2) + \cdots + 1$ multiplications,
$n - 1 + (n - 2) + \cdots + 1$ additions.
As we already noted

$$n - 1 + (n - 2) + \cdots + 1 = \frac{n(n-1)}{2}$$

The conclusion of all these calculations is thus the following:

The Gauss Method costs

$\frac{(n+1)n}{2} + n$ **divisions**

$\frac{n^3-n}{3} + \frac{n(n-1)}{2} + \frac{n(n-1)}{2} = \frac{n^3-n}{3} + n(n-1)$ **multiplications**

$\frac{n^3-n}{3} + \frac{n(n-1)}{2} + \frac{n(n-1)}{2} = \frac{n^3-n}{3} + n(n-1)$ **additions**

The most relevant summands continue to be $\frac{n^3}{3}$ and hence one concludes by saying that the **Gauss Method costs** $O(\frac{n^3}{3})$, by which we mean that the order of magnitude of the computational costs is $\frac{n^3}{3}$ operations.

Surely some readers will want to ask the precise meaning of the statement "the relevant part of the cost is $\frac{n^3}{3}$" or that "the order of magnitude of the cost is $\frac{n^3}{3}$". Let's see if we can satisfy such readers. First consider that it is of little importance what the cost is to solve a *small* system of linear equations, for example with $n = 2$ or $n = 3$ equations, because in such cases the cost is practically zero for whatever calculator we use.

But, having an idea of the number of operations that one performs, or more accurately that the calculator performs, becomes essential when n gets very big. For example, when $n = 100$ the total number of multiplications is

$$\frac{100^3 - 100}{3} + 100 \times 99 = 343,200$$

If we make a partial count we see that $\frac{100^3 - 100}{3} = 333,300$, and $100 \times 99 = 9900$ and that $\frac{100^3}{3} \cong 333,333$. The consequence of this calculation shows that $\frac{100^3}{3}$ is a pretty good approximation to the correct answer. Moreover $\frac{n^3}{3}$ is a better and better approximation as n grows i.e. as the type of the matrix grows. This is the reason why we say that the computational cost of the Gaussian Method is $\frac{n^3}{3}$.

Let's make another observation for the mathematically curious (I hope that there are such among the readers of this book). Consider the sequence $F(n) = \frac{n^3/3}{\text{cost}(n)}$, where $\text{cost}(n)$ is the number of multiplications needed to solve a square system of linear equations whose matrix is of type n. A mathematician observes that $\lim_{n \to \infty} F(n) = 1$ and this fact leads the mathematician to conclude that the two functions have the same *order of magnitude* and thus to say that the Gauss Method costs $O(\frac{n^3}{3})$.

Another important aspect of the calculations is the choice of the pivot. From a purely theoretical point of view the only important thing about the pivot is that of being *different from zero*. But, as we saw in the introductory chapter, *there are ways and there are ways of being different from zero*. All kidding aside, we will see in a little bit what can happen when one uses *approximate arithmetic* and chooses a pivot which is *very small*.

First, however, let's make an observation which has enormous importance in practical calculations. When we spoke of computational cost we always made the hypothesis that the cost of each single operation didn't depend on that operation. But, it's clear that such an assumption is valid only if each number entered has a finite and constant cost. In order for that to be the case we cannot move in a purely symbolic environment with integers or rational numbers where approximation is not permitted. In fact, it is clearly ridiculous to maintain that the cost of multiplying 2×3 is the same as that of multiplying $2323224503676442793 \times 3737625382643962983389217128$.

On the other hand, as we already noted in the introductory chapter, the use of approximate numbers may have disastrous consequences if we don't take adequate precautions. We are not going to get into this sort of problem here, but we'll show (with an example) how the choice of the pivots in Gaussian elimination requires special care if we use approximations.

Example 3.4.1. A small pivot
Let's consider the following system of linear equations

$$\begin{cases} 0.001x + y = 1 \\ x + y = 1.3 \end{cases} \tag{1}$$

Let

$$A = \begin{pmatrix} 0.001 & 1 \\ 1 & 1 \end{pmatrix} \qquad \mathbf{b} = \begin{pmatrix} 1 \\ 1.3 \end{pmatrix} \qquad \mathbf{x} = \begin{pmatrix} x \\ y \end{pmatrix}$$

and write the system as $A\mathbf{x} = \mathbf{b}$.

Let's not allow more than three digits after the decimal point and so we will use **round off** whenever there appear more than three digits after the decimal point. Recall that the only condition that a pivot has to satisfy is that of being different from zero, so we can use 0.001 as a pivot and the matrix then transforms into

$$A_2 = \begin{pmatrix} 0.001 & 1 \\ 0 & -999 \end{pmatrix} \qquad \mathbf{b}_2 = \begin{pmatrix} 1 \\ -998.7 \end{pmatrix}$$

The second equation provides us with the solution $y = \frac{9987}{9990}$. Exactly what is the value of $\frac{9987}{9990}$? In exact terms, the number is already correctly expressed as a fraction. Its decimal representation is $0.999(699)$. Rounding up the three digit decimal gives us 1 with an error of $1 - 0.999(699) \cong 0.0003$, i.e. of the order of *three ten thousandths*. We can go on to substitute $y = 1$ in the first equation and we obtain the equation $0.001x + 1 = 1$, from which we deduce that $x = 0$.

Now, interchange the two equations and proceed using 1 as the pivot. One gets

$$A = \begin{pmatrix} 1 & 1 \\ 0.001 & 1 \end{pmatrix} \qquad \mathbf{b} = \begin{pmatrix} 1.3 \\ 1 \end{pmatrix}$$

$$A_2 = \begin{pmatrix} 1 & 1 \\ 0 & 0.999 \end{pmatrix} \qquad \mathbf{b}_2 = \begin{pmatrix} 1.3 \\ 0.9987 \end{pmatrix}$$

The number 0.9987 is rounded up to 0.999 and so one obtains $y = 1$ which we substitute into the first equation and that gives us $x = 0.3$. We thus obtain a noteworthy discrepancy in the results. With the pivot 0.001 we obtained the solution $(0, 1)$, with the pivot 1 we obtained the solution $(0.3, 1)$. But, what is the exact solution? If we don't round up, in the second case we obtain the solution $y = \frac{0.9987}{0.999}$, and then substituting in the first equation we get $x = 1.3 - \frac{0.9987}{0.999}$. Thus, the exact solution is

$$\left(\frac{10039}{33330}, \frac{9987}{9990} \right)$$

Rounding off the result to three decimals we obtain the solution $(0.301, 1)$. The conclusion is that the second choice of the pivot brought us to a more reasonable result while the first one did not because in the first case the pivot we choose was small with respect to the other coefficients.

3.5 The *LU* Decomposition

It's interesting to study a kind of decomposition of square matrices called the *LU decomposition* or *LU form*. This plays an important role in the study of systems of linear equations especially (but not only) from the computational point of view.

First of all where do the letters L and U come from? You don't need a great deal of imagination to understand that they come from the words *Lower* and *Upper*. The name refers to the fact that one can decompose certain square invertible matrices as the product of a matrix L, i.e. *lower triangular*, with a matrix U, i.e. *upper triangular*. Let's begin with the observation that such a decomposition is **not** always possible.

Example 3.5.1. LU is not always possible

Let $A = \begin{pmatrix} 0 & 1 \\ 1 & 0 \end{pmatrix}$ and suppose that we have $A = LU$ with L a lower triangular matrix and U an upper triangular matrix. We thus have

$$L = \begin{pmatrix} \ell_{11} & 0 \\ \ell_{21} & \ell_{22} \end{pmatrix} \qquad U = \begin{pmatrix} u_{11} & u_{12} \\ 0 & u_{22} \end{pmatrix}$$

From the equality $A = LU$ one obtains

$$\ell_{11}u_{11} = 0, \quad \ell_{11}u_{12} = 1, \quad \ell_{21}u_{11} = 1, \quad \ell_{21}u_{12} + \ell_{22}u_{22} = 0$$

The first three equalities are incompatible because the second and the third force ℓ_{11} and u_{11} to be different from zero and that makes the first equation not solvable. The conclusion is that A doesn't have an LU decomposition.

Let's suppose that we have a square matrix A with the property that when we calculate the inverse the pivots can always be found *without using the operation of exchanging rows*.

Suppose that $E_1, E_2, \ldots, E_r$ are the elementary matrices which correspond to the elementary operations which transform A into a matrix U which is upper triangular, i.e. those operations that one does in the first part of the Gauss Method. One has

$$E_r E_{r-1} \cdots E_1 A = U \tag{$*$}$$

Given that the elementary matrices E_i correspond either to the product of a row by a constant or to the sum of a row with some *preceding* row multiplied by a constant, a moment of reflection will reveal that each matrix is lower triangular. We note that this is not true in Example 3.5.1, since, in that example we had to interchange rows to choose the first pivot. Now observe that from formula ($*$) we obtain

$$A = E_1^{-1} E_2^{-1} \cdots E_r^{-1} U \tag{$**$}$$

Mathematicians assure us that the following two facts hold.

(1) **The inverse of a lower triangular (upper triangular) matrix is a lower triangular (upper triangular) matrix.**

(2) The product of two lower triangular (upper triangular) matrices is lower triangular (upper triangular).

From these facts we can conclude that the matrix

$$L = E_1^{-1} E_2^{-1} \cdots E_r^{-1}$$

is lower triangular and thus the formula $(**)$ can be read as

$$A = LU$$

which is exactly what we wanted. The more attentive reader should not have any difficulty in proving the two facts we mentioned above. In fact, the second follows directly from the definition of the product of matrices, while the first can be proved following the reasoning we made in Section 3.3.

We conclude this section with some comments on the potential usefulness of the LU decomposition. Suppose we would like to solve a system of linear equations $A\mathbf{x} = \mathbf{b}$ with A invertible and assume we know how to write $A = LU$ as above. In this case it is less work to solve the system then to find the inverse of A, in the sense that the cost in terms of the number of operations is less. We proceed in the following way.

We write the system as $LU\mathbf{x} = \mathbf{b}$. If we put $U\mathbf{x} = \mathbf{y}$, then the original system becomes $L\mathbf{y} = \mathbf{b}$. First we solve $L\mathbf{y} = \mathbf{b}$ and obtain $\mathbf{y} = \mathbf{b}'$. Then we substitute and obtain $U\mathbf{x} = \mathbf{b}'$. Now it's enough to solve the system $U\mathbf{x} = \mathbf{b}'$ and that will give us the solutions to $A\mathbf{x} = \mathbf{b}$, Hold on a minute, in order to solve the original system of linear equations we had to solve two others! What kind of saving is that?

In fact, the two systems we had to solve had coefficient matrices that were triangular and thus, doing an analysis of the operations we need, like we did in Section 3.4 we see that the cost is *of the order of* $\frac{n^2}{2}$ *multiplications*, in marked contrast with $\frac{n^3}{3}$ in the general case. It's not hard to convince yourself that $2 \cdot \frac{n^2}{2}$ is of lesser order than $\frac{n^3}{3}$.

3.6 Gaussian Elimination for General Systems of Linear Equations

Not all systems of linear equations have as many equations as unknowns and even in that case, the coefficient matrix is not always invertible. We have arrived at the moment when we have to deal with the general problem of solving any system of linear equations. As usual, we begin by looking at an example.

Example 3.6.1. A non-square system of linear equations

Let's consider the following linear system

$$\begin{cases} x_1 + 2x_2 + 2x_3 \quad\quad + 7x_5 = 1 \\ -x_1 - 2x_2 - 4x_3 + x_4 - 2x_5 = 0 \\ x_1 + 2x_2 + 3x_3 \quad\quad + 4x_5 = 0 \end{cases} \tag{1}$$

If we set

$$A = \begin{pmatrix} 1 & 2 & 2 & 0 & 7 \\ -1 & -2 & -4 & 1 & -2 \\ 1 & 2 & 3 & 0 & 4 \end{pmatrix} \quad b = \begin{pmatrix} 1 \\ 0 \\ 0 \end{pmatrix} \quad x = \begin{pmatrix} x_1 \\ x_2 \\ x_3 \\ x_4 \\ x_5 \end{pmatrix}$$

we can write the system as $Ax = b$. Quite naturally we find ourselves with a non-square coefficient matrix A, just as we saw in Example 3.1.2. Recall that we had left that example up in the air in Section 3.1, although we could have acted upon it with elementary operations so as to *simplify it*. But, what would be the point?

Let's see. We'll do a few elementary operations on A, trying to be as methodical as possible. The reason for trying to be methodical is that we would like to eventually do this sort of thing with a calculator and for that reason we will want *to construct an algorithm*.

We know that we can replace the system (1) with an equivalent system that is obtained by replacing the second equation with the sum of the first and second equation. In this way we produce the matrices

$$A_2 = \begin{pmatrix} 1 & 2 & 2 & 0 & 7 \\ 0 & 0 & -2 & 1 & 5 \\ 1 & 2 & 3 & 0 & 4 \end{pmatrix} \quad b_2 = \begin{pmatrix} 1 \\ 1 \\ 0 \end{pmatrix}$$

Now we can replace the third equation with the third equation minus the first, thus obtaining the matrices

$$A_3 = \begin{pmatrix} 1 & 2 & 2 & 0 & 7 \\ 0 & 0 & -2 & 1 & 5 \\ 0 & 0 & 1 & 0 & -3 \end{pmatrix} \quad b_3 = \begin{pmatrix} 1 \\ 1 \\ -1 \end{pmatrix}$$

Now we exchange the third equation with the second, thus obtaining the matrices

$$A_4 = \begin{pmatrix} 1 & 2 & 2 & 0 & 7 \\ 0 & 0 & 1 & 0 & -3 \\ 0 & 0 & -2 & 1 & 5 \end{pmatrix} \quad b_4 = \begin{pmatrix} 1 \\ -1 \\ 1 \end{pmatrix}$$

Replacing the third equation with the third equation plus twice the second equation, we obtain the matrices

$$A_5 = \begin{pmatrix} 1 & 2 & 2 & 0 & 7 \\ 0 & 0 & 1 & 0 & -3 \\ 0 & 0 & 0 & 1 & -1 \end{pmatrix} \quad b_5 = \begin{pmatrix} 1 \\ -1 \\ -1 \end{pmatrix}$$

The original system (1) is thus equivalent to the system

$$
\begin{cases}
x_1 + 2x_2 + 2x_3 \quad + 7x_5 = \quad 1 \\
\qquad\qquad x_3 \quad - 3x_5 = -1 \\
\qquad\qquad x_4 - \quad x_5 = -1
\end{cases}
\tag{5}
$$

At this point we can solve the system (5) by reasoning in the following way. If we let x_5 vary freely, in other words we transform x_5 into a **parameter**, we can solve the last equation in exactly the same way we did using the Gauss Method on a square matrix. We can put $x_5 = t_1$ and obtain $x_4 = -1 + t_1$ from the third equation and $x_3 = -1 + 3t_1$ from the second. Substituting into the first equation we obtain $x_1 + 2x_2 = 1 - 2(-1 + 3t_1) - 7t_1 = 3 - 13t_1$. Now we can freely vary x_2 transforming it into a parameter. We put $x_2 = t_2$ and we obtain $x_1 = 3 - 13t_1 - 2t_2$.

We conclude by saying that the general solution of system (5) is

$$
(3 - 13t_1 - 2t_2, \; t_2, \; -1 + 3t_1, \; -1 + t_1, \; t_1)
\tag{$*$}
$$

We see that there are an infinite number of solutions depending on two parameters. We describe this by saying that there are **infinity to the power two solutions** (and we write ∞^2). In order to find a specific solution it is enough to fix values for the parameters. For example, for $t_1 = 1$ and $t_2 = 0$ we obtain the solution $(-10, 0, 2, 0, 1)$, while for $t_1 = 1$ and $t_2 = 3$ we get, instead, the solution $(-16, 3, 2, 0, 1)$.

This seems to be the right time to make an observation of primary importance. Having arrived at the equation $x_1 + 2x_2 = 3 - 13t_1$, we could have proceeded in an entirely different way. For example, we could have chosen x_1 as the parameter and we would have then obtained, as a general solution, the following

$$
(t_2, \; \frac{1}{2}(3 - 13t_1 - t_2), \; -1 + 3t_1, \; -1 + t_1, \; t_1)
\tag{$**$}
$$

The reader is invited to reflect on the fact that the two expressions $(*)$ and $(**)$, although different, represent the same set of 5-tuples.

This example well illustrates the fact that the choice of the **free variables** is not, in general, forced. However, there is something about them that doesn't change, and that is their **number**.

We have to be a bit careful in our way of saying things. If, for example, we had a system of linear equations, with coefficients in $\mathbb{Z}_2$, whose solutions depended on two parameters, then the *enormous number of infinity to the power 2* is nothing more than the more modest number 4. In this case, each parameter can only assume two values, namely 0 and 1, and hence the pair of parameters can only assume the four values $(0, 0), (0, 1), (1, 0), (1, 1)$. Thus, the phrase *infinity to the power 2* really sounds well only in the cases for which the number field from which we choose the coefficients is itself infinite.

Another important consideration is that even if there are more unknowns than equations, this does not mean that the system must have solutions, as the following example shows.

Example 3.6.2. Many unknowns but no solution

Consider the following system of linear equations

$$\begin{cases} x_1 + 2x_2 + 2x_3 + 7x_5 = 1 \\ -x_1 - 2x_2 - 4x_3 + x_4 - 2x_5 = 0 \\ -2x_3 + x_4 + 5x_5 = 0 \end{cases} \qquad (1)$$

Setting

$$A = \begin{pmatrix} 1 & 2 & 2 & 0 & 7 \\ -1 & -2 & -4 & 1 & -2 \\ 0 & 0 & -2 & 1 & 5 \end{pmatrix} \qquad b = \begin{pmatrix} 1 \\ 0 \\ 0 \end{pmatrix} \qquad x = \begin{pmatrix} x_1 \\ x_2 \\ x_3 \\ x_4 \\ x_5 \end{pmatrix}$$

we can write the system as $Ax = b$. Replacing the second equation with the sum of the first and second equations we produce the matrices

$$A_2 = \begin{pmatrix} 1 & 2 & 2 & 0 & 7 \\ 0 & 0 & -2 & 1 & 5 \\ 0 & 0 & -2 & 1 & 5 \end{pmatrix} \qquad b_2 = \begin{pmatrix} 1 \\ 2 \\ 0 \end{pmatrix}$$

If we now replace the third equation with the third equation minus the second, we obtain the matrices

$$A_3 = \begin{pmatrix} 1 & 2 & 2 & 0 & 7 \\ 0 & 0 & -2 & 1 & 5 \\ 0 & 0 & 0 & 0 & 0 \end{pmatrix} \qquad b_3 = \begin{pmatrix} 1 \\ 2 \\ -2 \end{pmatrix}$$

The third equation $0 = -2$ doesn't have any solutions and, as a consequence, the system (1), even though it has three equations and five unknowns, also has no solutions.

In the next section a very important number, associated to a square matrix, enters the scene. Whether this number is zero or not zero will furnish us with fundamental information.

3.7 Determinants

We have seen many aspects of the theory of matrices and we have dwelt a great deal on the importance of both the notion and calculation of the inverse of a matrix. We have also seen that not every square matrix has an inverse. It's a natural question to ask if there is any way of knowing whether or not a matrix has an inverse, without trying to calculate it. It would be nice if there were some *oracle* that could tell us *a priori* if a given matrix has an inverse or not.

Given that science doesn't put much trust in supernatural forces like oracles, we can ask if there is some function which can give the desired response. Let's try to figure out in a logical way what could possibly give us such information. Let's consider a general matrix of type 2, i.e.

$$A = \begin{pmatrix} a_{11} & a_{12} \\ a_{21} & a_{22} \end{pmatrix}$$

and look at the number

$$d = a_{11}a_{22} - a_{12}a_{21}$$

Let's separate things into two cases

(1) The first column of A is zero.
(2) The first column of A is not zero.

In the first case we have already observed that the matrix is not invertible and we see that $d = 0 \cdot a_{22} - a_{12} \cdot 0 = 0$.
In the second case we have two subcases

(2a) The element $a_{11} \neq 0$.
(2b) The element $a_{11} = 0$.

In case (2a) we can use a_{11} as a pivot and with an elementary operation transform the matrix into

$$A_2 = \begin{pmatrix} a_{11} & a_{12} \\ 0 & a_{22} - \frac{a_{21}}{a_{11}}a_{12} \end{pmatrix} = \begin{pmatrix} a_{11} & a_{12} \\ 0 & \frac{d}{a_{11}} \end{pmatrix}$$

If $d \neq 0$ the matrix A_2 is invertible because it is upper triangular with all the elements on the main diagonal not zero. Thus, A is also invertible. If, instead, $d = 0$, the matrix A_2 is not invertible because it has a zero row, and hence the matrix A is also not invertible.

In case (2b), $a_{21} \neq 0$ is forced and so we can exchange the rows and obtain

$$A_2 = \begin{pmatrix} a_{21} & a_{22} \\ a_{11} & a_{12} \end{pmatrix}$$

and now we can use a_{21} as a pivot and, with an elementary operation, transform the matrix into

$$A_3 = \begin{pmatrix} a_{21} & a_{22} \\ 0 & a_{12} - \frac{a_{11}}{a_{21}}a_{22} \end{pmatrix} = \begin{pmatrix} a_{21} & a_{22} \\ 0 & -\frac{d}{a_{21}} \end{pmatrix}$$

Now we can argue as in case (2a) and conclude that if $d \neq 0$ the matrix A_3 is invertible and hence the matrix A is also invertible. If, instead, $d = 0$ then the matrix A_3 is not invertible because it has a row of zeroes, and hence the matrix A is also not invertible. At this point we have exhausted all the possible cases and we find, in our hands, the following unexpected fact.

The matrix A is invertible if and only if $a_{11}a_{22} - a_{12}a_{21} \neq 0$

We have found the much sought after oracle! The number $a_{11}a_{22} - a_{12}a_{21}$ will decide if a type 2 matrix is invertible. If $a_{11}a_{22} - a_{12}a_{21} = 0$, then the matrix A is not invertible while if $a_{11}a_{22} - a_{12}a_{21} \neq 0$, then the matrix A is invertible.

That number *determines* whether or not a type 2 matrix is invertible and hence merits a name: it has come to be called the **determinant of A** and denoted with the symbol **det(A)**. Since every square matrix of type 2 has a determinant we can speak of the **determinant function**.

But what does this function really measure? What we have seen about the determinant up to this point is very important, but it is only part of the story. Up to now we have seen the significance of whether or not the determinant is zero or not zero. What about when this number is not zero, does the number itself have some significance? What about if we have a square matrix of type bigger than 2, does there exist a determinant for such a matrix? For the moment we will only respond to the second question. The answer to it is "yes" and in the case of a square matrix of type 3 one has

$$A = \begin{pmatrix} a_{11} & a_{12} & a_{13} \\ a_{21} & a_{22} & a_{23} \\ a_{31} & a_{32} & a_{33} \end{pmatrix}$$

$$\det(A) = a_{11}a_{22}a_{33} - a_{11}a_{23}a_{32} - a_{12}a_{21}a_{33} + a_{12}a_{23}a_{31} + a_{13}a_{21}a_{32} - a_{13}a_{22}a_{31}$$

How can we ever expect to remember such a formula? And where does it even come from? For the moment we will only make an observation. For example, one can regroup the terms in the above formula and rewrite it as

$$\det(A) = a_{11}(a_{22}a_{33} - a_{23}a_{32}) - a_{12}(a_{21}a_{33} - a_{23}a_{31}) + a_{13}(a_{21}a_{32} - a_{22}a_{31})$$

This is called the expansion of the determinant along the first row, in the sense that one can read it as the sum, with alternating signs, of the product of the elements of the first row by the *determinants* of three matrices of type 2. Moreover, the rule is simple to remember because the matrix whose determinant is to multiplied by a_{ij} is the matrix one obtains by crossing out the i-th row and j-th column of the matrix A.

Another interesting observation is that $\det(A)$ can be obtained by expanding along any row or column. For example, if we regroup the summands in a different way we get

$$\det(A) = a_{11}(a_{22}a_{33} - a_{23}a_{32}) - a_{21}(a_{12}a_{33} - a_{13}a_{32}) + a_{31}(a_{12}a_{23} - a_{13}a_{22})$$

One can define, in an analogous way, the determinant of any square matrix of whatever type. Mathematicians have developed a theory which characterizes the determinant function as the unique function which satisfies certain formal properties. We'll come back to that in detail in Section 4.6. For the moment we will have to be satisfied with knowing just a bit more about the determinant and the importance of it being zero or not zero, at least in the case of square matrices of type two. Some questions will still be left unanswered, such as where did the formula come from and what does the value of the determinant represent.

We are now ready to close this chapter and we are ending it with some questions. As has already been said, in life and hence also in science, there are more questions than answers. Fortunately, there will be some answers in the next chapter.

Exercises

Exercise 1. Solve the linear equation

$$2x_1 + x_2 + x_3 + x_4 - x_5 = 0$$

Exercise 2. Solve the linear system

$$\begin{cases} 2x_1 + x_2 = 0 \\ x_1 - x_2 = 0 \end{cases}$$

@ **Exercise 3.** Solve the system of linear equations

$$\begin{cases} x_1 + \frac{2}{5}x_2 + 5x_3 - \frac{12}{3}x_4 + 2x_5 - 4x_6 = 0 \\ \frac{1}{2}x_1 + 3x_2 - x_3 - 13x_4 + 12x_5 - 3x_6 = 1 \\ \frac{1}{2}x_1 - 11x_2 - 3x_3 + 13x_4 - \frac{7}{2}x_5 - 2x_6 = 8 \\ 6x_1 + \frac{2}{7}x_2 + \frac{1}{2}x_3 + 14x_4 + 7x_5 - 2x_6 = 0 \\ 13x_1 + x_2 + \frac{1}{4}x_3 - 2x_4 + 22x_5 - 13x_6 = 7 \\ 9x_1 + \frac{1}{7}x_2 + 12x_3 + 13x_4 - 7x_5 - 2x_6 = \frac{1}{2} \end{cases}$$

@ **Exercise 4.** Let $t \in \mathbb{Q}$ be a parameter and consider the parameterized system of linear equations

$$\begin{cases} x + \frac{2}{5}y + z = 0 \\ ty - \frac{2}{3}z = 0 \\ tx - \frac{8}{5}y + \frac{7}{3}z = 1 \end{cases}$$

in the unknowns x, y, z. Describe the solutions as the parameter t varies.

@ **Exercise 5.** Given the family of linear systems (in the unknowns x, y, z, w)

$$\begin{cases} x + ay + 2z + 3w = 0 \\ -by + 3z + 3w = 0 \\ z + w = -1 \end{cases}$$

describe the solutions as $a, b \in \mathbb{Q}$ vary.

@ **Exercise 6.** Let x, y, z, w be unknowns and consider the family of systems of linear equations

$$\begin{cases} x + 2y + z = 0 \\ ax + y + 2z + 2w = 0 \\ -y + 3z + 3w = 0 \\ z + 3w = 0 \end{cases}$$

(a) Describe the solutions to the linear system obtained by varying $a \in \mathbb{Q}$.
(b) Let A be the coefficient matrix associated to the given system. Find two matrices $B, U \in \text{Mat}_3(\mathbb{R})$ such that B is invertible and U is upper triangular and such that $BU = A$.
(c) Find the values of $a \in \mathbb{R}$ such that A is invertible.
(d) For those values of a for which A is invertible, find A^{-1}.

Exercise 7. Given the following matrix

$$A = \begin{pmatrix} 1 & 1 & -2 \\ \frac{1}{2} & -2 & 1 \\ 1 & 0 & \frac{2}{5} \end{pmatrix}$$

calculate the LU decomposition of A.

@ **Exercise 8.** Calculate the LU decomposition of the following matrix

$$\begin{pmatrix} 1 & 1 & 2 & 1 & 2 & 1 \\ 1 & 10 & -1 & 4 & -10 & 4 \\ 2 & -1 & 6 & -1 & 9 & -1 \\ 1 & 4 & -1 & 7 & -3 & 8 \\ 2 & -10 & 9 & -3 & 23 & -7 \\ 1 & 4 & -1 & 8 & -7 & 36 \end{pmatrix}$$

Exercise 9. *(Difficult)*
Prove that a linear system with coefficients in $\mathbb{Z}_2$ cannot have exactly 7 solutions.

Exercise 10. *(Difficult)*
(a) Prove that the LU decomposition of a square matrix is not unique.
(b) Prove that if one asks, in addition, that L have all of its diagonal elements equal to 1, then the LU decomposition (when it exists) is unique.

Exercise 11. Given the family of matrices

$$A_a = \begin{pmatrix} 0 & 2-a & 1 \\ a-1 & 1 & 0 \\ a & a & 0 \end{pmatrix}$$

(a) Determine the values of $a \in \mathbb{R}$ for which the matrix A_a is invertible.
(b) Are there values $a \in \mathbb{R}$ for which one can find the LU decomposition of A_a?

Exercise 12. Let A be a square matrix of type n. Prove that if A has $n^2 - n + 1$ zero entries, then A is not invertible.

Exercise 13. In this exercise we will suppose that the matrices are square, of type 2, with entries in $\mathbb{Z}_2$.

(a) How many matrices are there in $\mathrm{Mat}_2(\mathbb{Z}_2)$?

(b) How many matrices in $\mathrm{Mat}_2(\mathbb{Z}_2)$ have determinant different from zero?

Exercise 14. Find, if it exists, the inverse of the matrix

$$\begin{pmatrix} 1 & 2 & 3 & 4 & 5 \\ 6 & 7 & 8 & 9 & 10 \\ 11 & 12 & 13 & 14 & 15 \\ 16 & 17 & 18 & 19 & 20 \\ 21 & 22 & 23 & 24 & 25 \end{pmatrix}$$

Exercise 15. Consider the general matrix in $\mathrm{Mat}_2(\mathbb{R})$, i.e. the matrix

$$A = \begin{pmatrix} a_{11} & a_{12} \\ a_{21} & a_{22} \end{pmatrix}$$

Call d the determinant of A and suppose that $d \neq 0$. Verify the following equality

$$A^{-1} = \frac{1}{d} \begin{pmatrix} a_{22} & -a_{12} \\ -a_{21} & a_{11} \end{pmatrix}$$

4

Coordinate Systems

question: *How much is a dollar worth?*
answer: *since you haven't specified*
a coordinate system (Au, Ag),
it's impossible to give you an answer

Up till now we have spoken about algebraic objects, usually matrices and vectors. But, at the beginning of this book (see Section 1.2) we gave examples of vectors such as force, velocity, acceleration, coming from physics. It's clear that we've used the word *vector* to mean at least two distinct things. But how can physical or geometric objects have the same name as purely algebraic ones? In this chapter we will study the *how* and the *why* and we will discover an extraordinary activity of the mathematical arts: the construction of models not only of physical, biological and statistical objects but also of other mathematical objects. Said another way: mathematical entities, often created to be models of something else, can themselves be modeled inside mathematics.

This discussion is starting to get a bit too technical so let me use a simple observation to give you some idea of where we are going in this chapter. It is part of the cultural heritage of *everyone* that in order to measure something it is necessary to have some standard unit of measure. Saying that a pole has length 3 doesn't mean anything to anyone, while it's clear what is meant by saying that a pole is three feet long. To say that one city is at a distance 30 from another doesn't say much, but it's clear what is meant by saying that one city is at a distance of 30 kilometers from another. It is the lack of a standard unit of measurement that stops us from understanding the meaning of the numbers in the previous sentences.

There is, however, one enormous and absurd exception to this rule. From the moment that money lost its fixed value in gold and silver, no one has been able to respond to the question: What is a dollar (or a Euro, or a yen, or a...) worth? We thus live in an economic-financial world which lacks a standard reference system.

Robbiano L.: Linear Algebra for everyone
© Springer-Verlag Italia 2011

In order to alleviate the stress which has surely been generated by this last observation, I invite the reader to move quickly to the first section of this chapter. But, first allow me to make the following striking statement: in this chapter we will begin to see how to geometrize algebra and algebraicize geometry.

4.1 Scalars and Vectors

In Section 1.2 we saw examples of vectors which come from physics. Now, we return to those examples and examine, in detail, some everyday descriptions of them.

- At the point P on the wall, the temperature is 15 degrees centigrade.
- A ball moves at a velocity of 15 centimeters a second as it passes the point P on the table.
- The car was moved three meters.

We notice immediately that only the first sentence gives a complete description. The other two are ambiguous and, right away, you want to ask questions: in which direction? in what sense or orientation? In order to give a thorough response we can visualize the first situation in the following way:

For the second situation we could use a representation of the type

For the third situation we could use a representation of the form

The second and third cases represent phenomena with more *attributes*, i.e. they have not only a property representable by a number, but also one given by a pointed line segment and one given by an arrow. The temperature, we say, is a **scalar quantity** while the velocity and the orientation are called **vector quantities**.

It is important to note that the number, that is the scalar quantity present also in the vector quantity, could be represented in the length of the line segment. Such an oriented segment has come to be known as a **vector**. Every vector is thus characterized by a **direction**, an **orientation** and a **modulus** (*pl. moduli*) or length of the segment which represents it.

If I now say that the automobile was moved two meters along a given direction, with a specific orientation, then, knowing where it was previously we know where it is now. The phenomenon is thus completely described by a vector.

What's the difference between the second and third cases? In the second case I was careful to specify where I measured the velocity of the ball while in the third case I could have specified *any point* as the original location of the car. Without going into too many technical details we say that this difference distinguishes the concept of **based vector** from that of **free vector** . In other words, a free vector is nothing more than the collection of based vectors you get by displacing a particular based vector in a *parallel manner*. This is a particular case of an **equivalence relation**, a fundamental concept in mathematics. Mathematicians say that one has imposed a kind of equality on the set of based vectors, where classes of parallel based vectors, i.e. the free vectors, are equal.

Let's not get overwhelmed by the jargon and think that mathematicians like abstract and abstruse concepts because mathematicians are snobs! The relationship of equivalence that we just described precisely expresses that which was said more vaguely earlier, i.e. the idea that parallel vectors may represent the same thing. It's just that, in mathematics, one needs to be rigorous above all when one is discussing basic notions (which is exactly where we are right now). If we didn't do that, then as soon as we'd passed this initial phase, we'd be unable to proceed correctly and we would eventually be forced to go back and fix things up.

If, however, you really cannot stand the concept of free vectors, then we'll take away their freedom! In order to do that, we fix a point O and then every free vector can be represented by a vector based at the point O. This is the first important step in transforming geometry into algebra and thus permitting us to use algebraic techniques to solve geometric problems. However, other steps also have to be taken.

4.2 Cartesian Coordinates

In the previous section we have seen that free vectors can all be chained and forced to have the same base point.

Now suppose that we are interested in free vectors that can move only in one direction. If we think of them as all based at the same point O, we can restrict these vectors to living on a line.

O

So, if we want to represent a vector it's enough to say where to find the "not O" end of the vector, i.e. to find where A is.

Now, if we want to improve the usefulness of the line, we can give it a **unit of measure** which will allow us to measure the length of the vector, i.e. the length of the segment OA. However, there is still some ambiguity: if we say that A is at a distance of 5 units from O we won't know on which side of O to look for A. It's enough to add an **orientation** to the line and call that orientation positive. The line is then called an **oriented line** and we've completed our job.

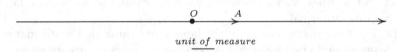

unit of measure

But, what "job" are we talking about? Observe that, with the properties our line now has, every free vector that has the direction of the line is represented by a vector based at O and hence is determined by its other end A. The length of the segment OA, measured with the unit of measure, is a real number to which we attach a positive sign if A is on the same side of the arrow which indicates the orientation of the line, negative if it's on the other side. It may not seem that we have done much, but, in fact, we've made a fundamental leap which now permits us to represent free vectors with a given direction, first, as points on a line, and then, as positive or negative real numbers.

In this way we have given birth to, what is called in mathematics, a **system of cartesian coordinates** on the line. What, then, is a system of cartesian coordinates on a line? It's a system constructed from: a line, a privileged point (called O) on that line and referred to as the **origin of the coordinate system**, a **unit of measure** which allows us to measure the lengths of segments of the line and thus the moduli of vectors, and an **orientation** that allows us to decide on which side of O the vector belongs.

For example, if we say that a vector on the line is represented by -5, what we want to mean is that we have the vector based at O and having as its other end a point at a distance 5 from O and which is on the side opposite the privileged orientation of the line.

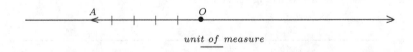

unit of measure

In this way we have succeeded in describing the whole class of free vectors using *just one real number*. This whole discussion can be further condensed. In fact, the orientation of the line and the unit of measure can both be described with one vector u_1 which is based at O and whose length is exactly that of the unit of measure. In this case, the vector u_1 is called a **unit vector**. Our system of coordinates on the line can be visualized in the following way

and we'll call it $\Sigma(O; u_1)$.

The fact that in the coordinate system $\Sigma(O; u_1)$ a given vector u has coordinate a_1 can be described using the equality $u = a_1 u_1$.

But what happens when the vectors are in the plane? The basic idea is to use *two oriented lines* which meet at a single point.

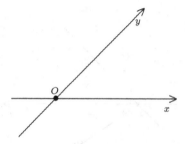

The two lines will be called the x-**axis** and the y-**axis** or sometimes simply the **coordinate axes**. Each of these axes has its own unit vector and, if we assume that the two lines are each given the same unit of measure then the coordinate system is called **monometric**.

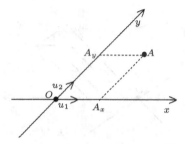

The system of coordinates is denoted $\Sigma(O; u_1, u_2)$. If we reason in the same way that we did in the case of a line, we see that every free vector can be represented by a vector based at O and thus is determined by its other end, A. Now, look at the point A and draw two lines through it, each parallel to an axis. We obtain two points A_x, A_y, as in the figure. Since the point 0 and the two unit vectors u_1 and u_2 give each of the axes a coordinate system, we can associate a real number to each of A_x and A_y. If those numbers are a_1, a_2 we will say that the coordinates of the vector OA (or of the point A) are (a_1, a_2). In this case, we will often write $A(a_1, a_2)$.

If we let u denote the vector OA we can also write

$$u = a_1 u_1 + a_2 u_2$$

(even if, for the moment, it's not clear what the symbol "+" in the equation means – we'll see what it means a little further on in Section 4.3).

The reasoning is similar in the case of space. Here we use *three directed lines which meet in one point and which are not, all three, in the same plane.*

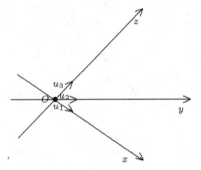

The three lines will be called the *x*-axis, *y*-**axis**, and *z*-**axis**. The (unique) plane which contains the x and y axes will be called the xy-**plane**, that which contains the x and z axes the xz-**plane** and that which contains the y and z axes, the yz-**plane**. These three planes are called the **coordinate planes**.

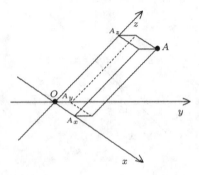

The coordinate system is denoted $\Sigma(O; u_1, u_2, u_3)$. Reasoning as earlier, every free vector can be represented as a vector based at O and thus is determined by its other end, A. We then consider the point A and from it trace the three planes which are parallel to the coordinate planes. We do this in such a way so as to obtain three points A_x, A_y, A_z, as in the figure. Given that the point O and the three unit vectors give a coordinate system on each of the axes, we may associate real numbers to the points A_x, A_y and A_z. If those real numbers are a_1, a_2, and a_3, we will say that the coordinates of the vector OA (or of the point A) are (a_1, a_2, a_3). In this case we often write $A(a_1, a_2, a_3)$.

As in the case of the line and the plane, *letting u denote the vector OA, one may also write $u = a_1 u_1 + a_2 u_2 + a_3 u_3$.* Shortly, (in Section 4.3) we'll see why we can do this.

At this point it is worthwhile to stop a moment and reflect on what we have done. In Section 1.2 we spoke of vectors and identified them with rows of matrices or columns of matrices. Essentially, when we spoke of vectors, we were thinking of them as ordered n-tuples of numbers. Furthermore, we said that vectors and matrices are often mathematical models for real problems and situations. We gave several examples.

What we have seen in this section can be interpreted in the same way. We again speak of vectors, but this time with a geometric twist, vectors are now represented by directed line segments on the line, in the plane or in space. Using the tool called "coordinate system" we can, in turn, represent such vectors using points and represent the points with real numbers (or pairs of real numbers or triples of real numbers). The correspondence, which we saw above, between free vectors on the line and real numbers, is called a **one to one correspondence** because one may make the correspondence in either way, i.e. from free vectors on the line to real numbers or from real numbers to free vectors on the line. This correspondence also gives us a one to one correspondence between real numbers and vectors based at O. The same discussion works analogously with free vectors in the plane and ordered pairs of real numbers and free vectors in space and ordered triples of real numbers.

But then, just exactly what is a system of cartesian coordinates, for example in the plane? It really is nothing more than a tool which allows us to identify free vectors (or vectors all based at the same point) with pairs of real numbers. Used in the opposite sense, it is a tool which allows us to identify ordered pairs of real numbers with free vectors in the plane.

Thus, a coordinate system in the plane (and analogously on the line and in space) allows us to use geometry to visualize, by means of vectors, ordered pairs of real numbers. In fact, these are the same pairs of real numbers which, in Section 1.2, we called vectors. Finally we have freed ourselves of a considerable ambiguity. Now we understand better the *double use of the word vector*!

A number, a pair of numbers, a triple of numbers, no matter what the situation or the problem that presents us with them, can be visualized as vectors. The benefit to us is that we can use geometric facts or geometric intuition to think about them. This realization constitutes one of the great advances in mathematics. Let's see how this works in practice: for every equation with two unknowns we can consider the set of all of its solutions. If we have fixed a coordinate system in the plane then that set of solutions corresponds to a set of points in the plane and thus to a geometric figure. This realization is a fundamental step in the geometrization of algebra (as promised in the introduction).

At this point new, very natural, questions come up. The main one is: what are the benefits of this theory? All this effort would seem excessive if, at best, this whole geometric discussion allows us only to deal with triples of real numbers. Will this ever help us, for example, handle the case of the 13-tuples that came up in the example of the "football pool" (see Example 1.2.2)?

We will look for answers to these questions in what follows, but first let's fix a bit of terminology. *The real numbers will be denoted with* $\mathbb{R}$, *ordered pairs of real numbers with* $\mathbb{R}^2$ *and ordered triples of real numbers with* $\mathbb{R}^3$. The geometric intuition which coordinate systems gave us, stops with the triples! But, using the algebraic formalization there's no problem in considering after the triples, quadruples, quintuplets, sextuplets and so on. In what follows we will freely construct and use the sets $\mathbb{R}^n$ for any natural number, n.

4.3 The Parallelogram Rule

In this section we begin to harvest some of the benefits we spoke about in the last section. The *geometrization* of vectors gives us geometric ways of looking at things which we had discussed previously only in purely algebraic ways.

First we must establish some notation. Given a coordinate system with origin O then, if u is the free vector represented by the vector based at O and having other end A then, from now on, we will usually write

$$u = A - O$$

The reason for using this strange symbol $A - O$ to represent the vector will be clear in a moment. But first I would like to linger a bit on the *mathematical audacity* of what we have written! How can a free vector equal a based vector? In fact, they cannot be the same and this way of writing things is a clear *abuse of notation*. One could correctly write $A - O \in u$ or $u \ni A - O$, since in fact these both have the following meaning: $A - O$ *is in the class of the free vector* u. One uses the equality symbol but the real meaning is that of "belonging". To make an analogy with everyday things it would be like calling a person who comes from Montreal "a Montreal", instead of a Montrealer.

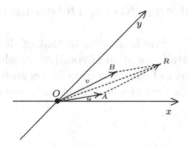

If we accept this abuse of notation then we can also accept writing $u = R - B$ to express the fact that u is the free vector represented by the vector based at B and having its other end at R. An important understanding comes from looking carefully at the figure above. Let's "read" it in the following way: suppose we are given a cartesian coordinate system (we haven't drawn the unit vectors on each axis in order not to clutter the diagram) and in it two free vectors, represented as $u = A - O$ and $v = B - O$. Suppose that the coordinates of A and B are, respectively, (a_1, a_2) and (b_1, b_2). The point R, then, is that unique point so that the $OARB$ is a parallelogram.

And now we notice immediately that $u = R - B$, $v = R - A$, and as a consequence, the coordinates of the point R are $(a_1 + b_1, a_2 + b_2)$.

The algebraic evidence and the experimental convenience induced us to define the sum of vectors and of matrices in the most natural way possible, i.e. that of summing the entries term by term. This sum now has a geometric interpretation as well. Having used a coordinate system to identify vectors with pairs of real numbers, the sum made by adding the coordinates of the vectors, term by term, corresponds to the sum of vectors made according to, what is called, the **parallelogram rule** as illustrated in the figure above. We have achieved our first important goal. I never get tired of repeating that, as we have just seen, using cartesian coordinate systems opens the door to the algebraization of geometry and to the geometrization of algebra.

Another observation follows from the considerations above. Returning to our figure, we said that $v = B - O = R - A$ and that made us reflect upon the seemingly careless use of the equality symbol.

But, if we consider the coordinates of the points in play, we observe that there are honest equalities in the expressions $(b_1, b_2) - (0, 0) = (b_1, b_2)$ and $(a_1 + b_1, a_2 + b_2) - (a_1, a_2) = (b_1, b_2)$. This explains the utility of the symbol $R - A$ to represent both the vector based at A and having other extreme at the point R, with the free vector associated to it. The parallelogram rule which we have just seen, guarantees that the coordinatesof such free vectors are exactly the differences of the coordinates of R and A.

4.4 Orthogonal Systems, Areas, Determinants

Now we will consider a situation analogous to that of the previous section, but in the context of a cartesian system of coordinates which is monometric and orthogonal, i.e. a system of cartesian coordinates such that the axes are pairwise orthogonal and the unit of measure on each axis is the same.

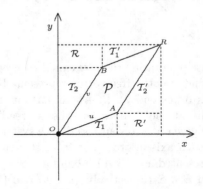

If the coordinates of A are (a_1, a_2) and the coordinates of B are (b_1, b_2) then, as we have already observed and discussed in the previous section, the coordinates of the point R are $(a_1 + b_1, a_2 + b_2)$, thus the area of the rectangle with two sides on the coordinate axes, determined by the points O and R is $(a_1 + b_1)(a_2 + b_2)$. Now we want to calculate the area of the parallelogram $\mathcal{P}$, i.e. the parallelogram defined by u and v. We observe immediately that the two rectangles $\mathcal{R}$, $\mathcal{R}'$ have equal areas, as do the two triangles $\mathcal{T}_1$, $\mathcal{T}_1'$ and also $\mathcal{T}_2$, $\mathcal{T}_2'$. A simple calculation shows

$$\text{Area}(\mathcal{P}) = (a_1 + b_1)(a_2 + b_2) - 2\,\text{Area}(\mathcal{R}) - 2\,\text{Area}(\mathcal{T}_1) - 2\,\text{Area}(\mathcal{T}_2)$$

and thus

$$\text{Area}(\mathcal{P}) = (a_1 + b_1)(a_2 + b_2) - 2a_2b_1 - a_1a_2 - b_1b_2 = a_1b_2 - a_2b_1$$

It seems like magic but it's true! We are dealing with a determinant! In fact, we can write

$$\text{Area}(\mathcal{P}) = \det \begin{pmatrix} a_1 & b_1 \\ a_2 & b_2 \end{pmatrix}$$

We have given a geometric interpretation to the concept of the determinant of a 2×2 square matrix whose entries are real numbers. If we read the two columns of the matrix as the coordinates of two vectors in a cartesian coordinate system with orthogonal and monometric axes, we get the following rule.

The absolute value of the determinant of a 2×2 square matrix with real entries coincides with the area of the parallelogram defined by the two vectors whose coordinates, in an orthogonal monometric cartesian coordinate system for the plane, form the columns of the matrix.

This answers a question which we left in the air at the end of the last chapter. As the attentive reader will have noticed, a small observation still remains before we can close this discussion. Area is, by its very nature, described by a non-negative number, while the determinant can be negative. This is why the absolute value of the determinant is the area mentioned above. The sign of the determinant indicates the orientation of the angle formed by the first and second vector. The sign is positive if the direction that one traverses, going from the first column to the second, is counterclockwise, negative if the direction is clockwise. We'll have a bit more to say about this in Section 6.2. In a perfect analogy with the case of the plane, one has the following rule.

The absolute value of the determinant of a 3×3 matrix with real entries coincides with the volume of the parallelepiped defined by the three vectors whose coordinates, in an orthogonal monometric cartesian coordinate system, form the columns of the matrix.

4.5 Angles, Moduli, Scalar Products

We've had success in giving the determinant a geometric significance and now we would like to continue on the path we have taken and push ahead with the geometrization of some algebraic concepts and the algebraization of some geometric concepts.

Keep in mind that in this section we will always work with an orthogonal monometric cartesian coordinate system.

A very natural question is the following: what is the length of a vector? And if the vector is represented in our coordinate system as $u = (a, b)$ can we find the length of u in terms of a and b? This is, like all questions, easy to ask – but fortunately, this time the answer is also easy to find.

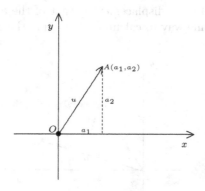

It's enough to use Pythagoras' Theorem to deduce that the length of u is $\sqrt{a_1^2 + a_2^2}$. We will indicate the length of a vector u by the symbol $|u|$, which will be called the **modulus** of u. Thus, if $u = (a_1, a_2)$ one has

$$|u| = \sqrt{a_1^2 + a_2^2}$$

In space we use exactly the same reasoning and if $u = (a_1, a_2, a_3)$ we get

$$|u| = \sqrt{a_1^2 + a_2^2 + a_3^2}$$

It's clear that the length of a vector does not depend on which based vector is chosen from the class of a free vector. Mathematicians would say that the length is an *invariant of the equivalence class*, a sure way to scare non-experts! However, an invariant is precisely that and, in this case, the invariance is based on empirical considerations, i.e. moving a vector to a position parallel to itself doesn't change its length (Einstein permitting!). In particular, if $u = B - A$ and if $A = (a_1, a_2)$, $B = (b_1, b_2)$, then

$$|u| = \sqrt{(b_1 - a_1)^2 + (b_2 - a_2)^2}$$

This formula may also be read as a formula describing the **distance** between the points A, B in the plane.

If $u = B - A$ and if $A = (a_1, a_2, a_3)$, $B = (b_1, b_2, b_3)$, then

$$|u| = \sqrt{(b_1 - a_1)^2 + (b_2 - a_2)^2 + (b_3 - a_3)^2}$$

which can be read as a distance formula between points A, B in space.

If a vector u is not the zero vector than there is a well defined vector which has the same direction but has unit length. That vector is called **the unit vector in the direction of** u or the **normalization of** u and is denoted by normal(u).

From the definition we get the formula

$$\text{normal}(u) = \frac{u}{|u|}$$

We're on a roll now and we don't want to stop yet. Another property which is invariant under parallel displacement is that of the **angle between two vectors**. Is there some way to calculate that using the coordinates only?

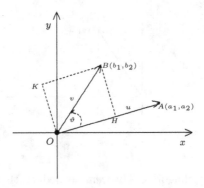

In order to solve this problem we ought to use an idea and formulas already well known to mathematicians. This is analogous to what we did for determinants, which were first defined algebraically, with their geometric nature being only later revealed. In this instance I am asking that the reader make an extra effort so he or she can better understand how mathematical reasoning works.

Let's remember that our problem is to express the angle between two vectors using their coordinates. In particular, given an orthogonal monometric cartesian coordinate system, we would like to be able to decide, knowing only their coordinates, if two vectors are orthogonal. What follows is typical mathematical reasoning.

> Let's suppose we have an orthogonal monometric system of cartesian coordinates and also two free vectors u and v in the plane (the reasoning in space is completely analogous). We draw them as the vectors $A - O$ and $B - O$ (see the figure above). From B we drop a perpendicular to the line through u and obtain the point of projection H. We choose the point K so that $OHBK$ is a rectangle. Let's remember that what we are after is a function φ which associates to every pair of vectors a real number. What sort of properties should the function φ have? If $\varphi(u, v)$ is supposed to somehow describe the angle between the two vectors formed by u and v, then we could require the following:

(a) If u, v are orthogonal, then $\varphi(u, v) = 0$.

We understand right away that this restriction on φ is not enough. Looking at the figure above, we observe that

$$H - O = (|v| \cos(\vartheta)) \, \mathrm{normal}(u)$$

and thus, in some way, we should bring the cosine of the angle ϑ into play. On the other hand, the figure shows that we have the equality

$$v = (H - O) + (K - O)$$

Let's suppose that our function has the following property.

(b) If $v = v_1 + v_2$, then $\varphi(u, v) = \varphi(u, v_1) + \varphi(u, v_2)$.

If that were the case, we could deduce the equality

$$\varphi(u, v) = \varphi(u, H - O) + \varphi(u, K - O)$$

But, the second summand is 0, by the property (a) and thus one has

$$\varphi(u, v) = \varphi(u, H - O) = \varphi(|u| \, \mathrm{normal}(u), |v| \cos(\vartheta) \, \mathrm{normal}(u))$$

Let's now suppose that our function has yet another property.

(c) If $u = cu'$, then $\varphi(u, v) = c \, \varphi(u', v)$ and analogously if $v = dv'$, then $\varphi(u, v) = d \, \varphi(u, v')$.

In that case, one would have the equality

$$\varphi(|u| \text{ normal}(u), |v| \cos(\vartheta) \text{ normal}(u)) = $$
$$|u||v| \cos(\vartheta) \varphi(\text{normal}(u), \text{normal}(u))$$

Let's also suppose that our function has the following property

(d) $\varphi(u, u) = |u|^2$.

Then, one would have $\varphi(\text{normal}(u), \text{normal}(u)) = 1$ and hence the equality

$$\varphi(u, v) = |u||v| \cos(\vartheta)$$

The above mentioned formula $\varphi(u, v) = |u||v| \cos(\vartheta)$ completely satisfies the requirement of giving information on the angle between u and v. But, we asked for a function $\varphi(u, v)$ that expressed this information using the coordinates of u and v. It looks as if we have not made any progress, but indeed we have. *This is not the time to give up!* Now, in fact, the problem has become the following: does there exist a function of the coordinates of the two vectors which satisfies all the properties we have specified?

Notice one subtle point: it will be enough to prove that there is a function; the uniqueness of such a function comes free of charge, given that, whatever the formulation of the function we give, it has to satisfy $\varphi(u, v) = |u||v| \cos(\vartheta)$. Thus the problem is simply to examine very carefully the requirements we put on our function. Let's look at those again.

(1) If u, v are orthogonal, then $\varphi(u, v) = 0$.
(2) The function $\varphi(u, v)$ should be linear both in u and in v.
(3) $\varphi(u, u) = |u|^2$

Condition (2) (we'll see, a bit later on, what this condition means) absorbs the conditions (b) and (c) which we had introduced earlier. We are finally ready to bring the coordinates into play. If, in our monometric, orthogonal cartesian coordinate system one has the equality $u = (a_1, a_2)$, $v = (b_1, b_2)$, then we obtain

$$|u+v|^2 = (a_1+b_1)^2 + (a_2+b_2)^2 = (a_1^2+a_2^2) + (b_1^2+b_2^2) + 2(a_1b_1+a_2b_2)$$

In other words

$$|u + v|^2 = |u|^2 + |v|^2 + 2(a_1b_1 + a_2b_2)$$

On the other hand, Pythagoras' Theorem says that $|u + v|^2 = |u|^2 + |v|^2$ if and only if u, v are orthogonal. Thus, the quantity $a_1b_1 + a_2b_2$ immediately becomes relevant. It is a function of the coordinates which, when it takes on the value 0, tells us that the two vectors are orthogonal i.e. satisfies property (1). Let's see if we have at hand the function we have been searching for. We put

$$\varphi(u, v) = a_1b_1 + a_2b_2$$

Now the road is easy; we are going downhill. This function, which satisfies property (1), satisfies (2) and (3). Let's check, for example, (3). If $u = (a_1, a_2)$ then

$$\varphi(u, u) = a_1 a_1 + a_2 a_2 = a_1^2 + a_2^2 = |u|^2$$

It is equally easy to verify (2). This then solves our problem.

The reasoning above has brought us to a solution and has also shown us that there is no other solution. The importance of the function we found is, therefore, enormous and to emphasize that we give it a particular name and a particular symbol.

If you haven't followed all the details of the discussion above, don't worry. What must be absolutely clear, however, is that we have defined a function that, given two vectors $u = (a_1, a_2)$, $v = (b_1, b_2)$, provides us with a number, namely $a_1 b_1 + a_2 b_2$. This function we indicate by writing $u \cdot v$ and we say it is the **scalar product** (sometimes **dot product**) of u and v. We have the formula

$$u \cdot v = a_1 b_1 + a_2 b_2 = |u||v| \cos(\vartheta) \tag{$*$}$$

Notice that, using this formula for the scalar product, we can also deduce the values of $|u|$ and $|v|$, in fact

$$|u| = \sqrt{u \cdot u}$$

In the case of space one proceeds in exactly the same way to obtain the analogous formula: if $u = (a_1, a_2, a_3)$, and $v = (b_1, b_2, b_3)$, then

$$u \cdot v = a_1 b_1 + a_2 b_2 + a_3 b_3$$

Also in this case $u \cdot v$ is called the scalar product of u and v. The formula analogous to $(*)$ is also valid, namely

$$u \cdot v = a_1 b_1 + a_2 b_2 + a_3 b_3 = |u||v| \cos(\vartheta)$$

In this section the reader has seen typical mathematical reasoning at work. I invite the reader, even one not very interested in mathematics, to meditate a bit on this system of formal logic. In a world rich with uncertainty it is helpful to recognize the strength of certain logical aspects of human thought.

4.6 Scalar Products and Determinants in General

Having arrived at this point, it's necessary to pause and formulate some general ideas regarding scalar products and determinants, both of which are fundamental mathematical concepts. We have to say that originally both were interpreted geometrically, but naturally only in the cases of vectors having two or three components. If we look at the definition of the scalar product, however, we see immediately that it is possible to forget the motivations and geometric interpretations which were useful for geometrizing $\mathbb{R}^2$ and $\mathbb{R}^3$, and freely move to any $\mathbb{R}^n$.

In fact, if $u = (a_1, a_2, \ldots, a_n)$ and $v = (b_1, b_2, \ldots, b_n)$ then it's not difficult to generalize the definition already given and write

$$u \cdot v = a_1 b_1 + a_2 b_2 + \cdots + a_n b_n$$

calling it still the scalar product. What are the most relevant properties of the scalar product?

(1) **Symmetry**, i.e. $u \cdot v = v \cdot u$ for every u and v.
(2) **Bilinearity**, i.e. linearity in both coordinates
$(a_1 u_1 + a_2 u_2) \cdot v = a_1(u_1 \cdot v) + a_2(u_2 \cdot v)$
$u \cdot (b_1 v_1 + b_2 v_2) = b_1(u \cdot v_1) + b_2(u \cdot v_2)$.
(3) **Positivity**, i.e. $u \cdot u = |u|^2 \geq 0$ and $u \cdot u = 0$ if and only if $u = 0$.

There is a fundamental relation, which we won't prove here, called the **Cauchy-Schwarz inequality**, which says that

$$|u \cdot v| \leq |u||v|$$

and from which it follows, in the case that both of the vectors u, v are not zero, that

$$-1 \leq \frac{u \cdot v}{|u||v|} \leq 1$$

This inequality permits us to read $\frac{u \cdot v}{|u||v|}$ as the *cosine of an angle*! The frigid abstraction of the Cauchy-Schwarz inequality allows us to consider angles between two vectors which live in non-physical spaces like $\mathbb{R}^n$, with n as big as we like. In particular we can now speak of the **orthogonality of two vectors in $\mathbb{R}^n$**, in the sense that we say that two vectors u and v are orthogonal if $u \cdot v = 0$. For example, the two vectors $u = (1, -1, -1, 0, 3)$ and $v = (-1, 0, -1, 5, 0)$ of $\mathbb{R}^5$ are orthogonal. And we don't even have any need for systems of coordinates, given that the scalar product is a function of the components of the two vectors. Mathematicians can really go quite far with a bit of mathematical fantasy coupled with some logical capacity. Moreover, the most noteworthy thing about all of this abstraction is that it is immediately useful for important real world applications. For example, modern statistics is permeated by these concepts.

It's a bit more difficult to generalize the concept of determinant to the case of a square matrix of any size. In what follows we will give the definition and the principal properties without much comment, except for the important comment that (as in the case of the scalar product) the properties determine the definition in a unique way. This is yet another aspect of the harmony which is pervasive in the most important mathematical constructions.

Let's recall that a **permutation** of the natural numbers $\{1, 2, \ldots, n\}$ is an arrangement of those n numbers i.e. what mathematicians would describe as a 1-1 correspondence between the set $\{1, 2, \ldots, n\}$ and itself. For example, all the permutations of the set $(1, 2, 3)$ are

$$(1, 2, 3), \; (1, 3, 2), \; (2, 1, 3), \; (2, 3, 1), \; (3, 1, 2), \; (3, 2, 1)$$

You can see right away that the number of permutations of $\{1, 2, \ldots, n\}$ is precisely $n \cdot (n-1) \cdots 2 \cdot 1$, i.e. the product of the first n natural numbers (this number is usually written as $n!$ and called n **factorial**).

If π is the name of a permutation of $(1, 2, \ldots, n)$, then π is usually written as $(\pi(1), \pi(2), \ldots, \pi(n))$. The sign of the permutation π is by definition either $+$ or $-$, according to whether the number of interchanges necessary to bring $(\pi(1), \pi(2), \ldots, \pi(n))$ into their natural order is even or odd. The operation of bringing the numbers $(\pi(1), \pi(2), \ldots, \pi(n))$ into their natural order can be done with different choices of interchanges; however, one can prove that the parity of the number of interchanges is invariant. The reader is invited to try this on an example in order to convince himself or herself of this fact.

For example, the sign of $(1, 3, 2)$ is $-$, because it is enough to exchange the 3 with the 2 in order to obtain $(1, 2, 3)$; on the other hand the sign of $(3, 1, 2)$ is $+$, because with two interchanges one obtains $(1, 2, 3)$ (do you see at least two different ways to bring this permutation into its natural order?).

Now let $A = (a_{ij})$ be any square matrix of size $n \times n$. For each permutation $\pi = (\pi(1), \pi(2), \ldots, \pi(n))$ of $\{1, 2, \ldots, n\}$ consider the product $a_{1\pi(1)} a_{2\pi(2)} \cdots a_{n\pi(n)}$ multiplied by $+1$ or -1, depending on the sign of the permutation π. There is no standard terminology for this product but we will call this number a **derived product** of A. The sum of all the $n!$ derived products of A (one for each permutation of $\{1, 2, \ldots, n\}$) is the number $\det(A)$, which is called the **determinant** of A.

One can easily check, for example, that the determinant of a matrix of size 2×2 or of size 3×3 is exactly the same as the definition we gave in Section 3.7.

As you can imagine, the cost of calculating the determinant, using just the definition, would be prohibitive. For example, to use the definition to calculate the determinant of a 20×20 matrix, you would have to calculate $20!$ derived products. But, $20! = 2,432,902,008,176,640,000$, i.e. around two thousand five hundred million billions! (the exclamation mark here doesn't mean factorial just shock at how fast these numbers grow). Even if the life of the planet depended on it we would never be able to calculate that determinant using just the definition.

But we will see that the definition, accompanied by a bit of reasoning, will let us discover notable properties of the determinant function with which we will do great things. Below I'll list some of those properties but I will skip the proofs as they require varying degrees of ability. I'll leave it to the reader to choose, according to his or her taste, which of these properties to try and prove. I hope that the results are encouraging.

(a) **If we interchange two rows or two columns of the matrix A we only change the sign of the determinant. Thus, if A has** two rows or two columns which are equal then the determinant has to be zero.

(b) If a single row or single column of the matrix A is multiplied by a constant c, then the determinant is multiplied by c.

(c) If we add to a row of the matrix A a multiple of another row then the determinant doesn't change. The same thing is true for the columns.

(d) If A is an upper or lower triangular matrix then

$$\det(A) = a_{11}a_{22}\cdots a_{nn}$$

In particular, $\det(I_n) = 1$

(e) We have $\det(A^{\mathrm{tr}}) = \det(A)$.

(f) If A and B are two square matrices of size n, then

$$\det(AB) = \det(A)\,\det(B)$$

This last is called **Binet's Theorem**. Its proof is not easy.

(g) A square matrix with entries in some field of numbers (for example $\mathbb{Q}$ or $\mathbb{R}$) is invertible if and only if its determinant is different from zero.

We note that, as happened in the process of solving linear equations, the properties mentioned above permit us to perform the operations which will reduce a square matrix into one in triangular form *keeping track, at the same time, of changes to the determinant*. Then we can apply rule (d) above.

We note with satisfaction that if we return to the case of a matrix of size 20×20 and calculate the cost of reducing the matrix to triangular form, one sees that the total cost is on the order of $\frac{20^3}{3}$ multiplications, which is around $3,000$ *a good deal less that* 20! (in this case the exclamation point does signify factorial and also... satisfaction because we have gotten around a serious problem). Let's see a concrete example where we apply the observations above.

Example 4.6.1. Let's consider the following matrix with rational numbers as entries

$$A = \begin{pmatrix} 1 & 2 & 1 & 5 & 1 \\ \frac{1}{2} & 2 & 1 & -1 & 2 \\ 3 & 2 & \frac{2}{3} & 1 & 1 \\ 1 & 1 & 2 & 1 & 2 \\ 2 & 2 & 3 & 3 & 4 \end{pmatrix}$$

Let's do some calculations. With only elementary transformations of the type that don't change the determinant, we can transform the matrix A into the matrix

$$A_2 = \begin{pmatrix} 1 & 2 & 1 & 5 & 1 \\ 0 & 1 & \frac{1}{2} & -\frac{7}{2} & \frac{3}{2} \\ 0 & -4 & -\frac{7}{3} & -14 & -2 \\ 0 & -1 & 1 & -4 & 1 \\ 0 & -2 & 1 & -7 & 2 \end{pmatrix}$$

then into the matrix

$$A_3 = \begin{pmatrix} 1 & 2 & 1 & 5 & 1 \\ 0 & 1 & \frac{1}{2} & -\frac{7}{2} & \frac{3}{2} \\ 0 & 0 & -\frac{1}{3} & -28 & 0 \\ 0 & 0 & \frac{3}{2} & -\frac{15}{2} & \frac{5}{2} \\ 0 & 0 & 2 & -14 & 5 \end{pmatrix}$$

and now into the matrix

$$A_4 = \begin{pmatrix} 1 & 2 & 1 & 5 & 1 \\ 0 & 1 & \frac{1}{2} & -\frac{7}{2} & \frac{3}{2} \\ 0 & 0 & -\frac{1}{3} & -28 & 4 \\ 0 & 0 & 0 & -\frac{267}{2} & \frac{41}{2} \\ 0 & 0 & 0 & -182 & 29 \end{pmatrix}$$

and finally into the matrix

$$A_5 = \begin{pmatrix} 1 & 2 & 1 & 5 & 1 \\ 0 & 1 & \frac{1}{2} & -\frac{7}{2} & \frac{3}{2} \\ 0 & 0 & -\frac{1}{3} & -28 & 4 \\ 0 & 0 & 0 & -\frac{267}{2} & \frac{3}{2} \\ 0 & 0 & 0 & 0 & \frac{281}{267} \end{pmatrix}$$

Using the various rules we listed above, we know that $\det(A) = \det(A_5)$. But, A_5 is a triangular matrix and so we can use rule (d). We multiply together the elements of the diagonal and conclude that the $\det(A) = \frac{281}{6}$.

4.7 Change of Coordinates

Let's turn, for a moment, to the case of coordinates in the plane which we will use to motivate the more general situation of coordinates in $\mathbb{R}^n$. The question that we ask ourselves is the following: what happens when we are dealing with two distinct coordinate systems? More specifically, what is the relationship between the coordinates of the same vector in the two systems? Now, suppose that we are dealing with two systems $\Sigma(O; u_1, u_2)$ and $\Sigma(P; v_1, v_2)$. First of all we observe that we can break the problem down into two simpler problems, using the ancient, but always efficient strategy of *divide and conquer.* First we compare the system $\Sigma(O; u_1, u_2)$ with $\Sigma(P; u_1, u_2)$ and then $\Sigma(P; u_1, u_2)$ with $\Sigma(P; v_1, v_2)$. In the first comparison we have a system and its translation, which we can visualize using the following figure.

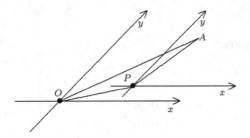

The parallelogram rule tells us

$$A - O = (A - P) + (P - O)$$

If the coordinates of A with respect to the system $\Sigma(O; u_1, u_2)$ are (a_1, a_2) and with respect to $\Sigma(P; u_1, u_2)$ are (b_1, b_2) and if the coordinates of P with respect to $\Sigma(O; u_1, u_2)$ are (c_1, c_2), then one has

$$\begin{pmatrix} a_1 \\ a_2 \end{pmatrix} = \begin{pmatrix} b_1 \\ b_2 \end{pmatrix} + \begin{pmatrix} c_1 \\ c_2 \end{pmatrix} \tag{$*$}$$

More interesting, but less obvious, is the second comparison. Now we have to come to grips with the problem of *change of basis*, i.e. where the origin of the coordinate systems is the same but both the directions of the coordinate lines and the size of the unit vectors may have changed.

In order to deal with this problem it will be good to establish notation which will be appropriate for the problem at hand and also suitable for future generalizations. Once we have said that the origin, P, of the two coordinate systems is the same, we then have to compare the coordinates of a vector with respect to $\Sigma(P; u_1, u_2)$ with the coordinates of the same vector with respect to $\Sigma(P; v_1, v_2)$. We'll indicate with $F = (u_1, u_2)$ the pair of vectors of unit length of the first system and with $G = (v_1, v_2)$ the pair of vectors of unit length of the second system.

We have already observed that if a vector v has coordinates (a_1, a_2) with respect to the first system, then this can be interpreted by writing

$$v = a_1 u_1 + a_2 u_2 \tag{1}$$

Now comes the good idea. The quantity $a_1 u_1 + a_2 u_2$ can be interpreted as the *row by column product* of F with $\begin{pmatrix} a_1 \\ a_2 \end{pmatrix}$. This fact suggests that we use the symbol

$$M_v^F = \begin{pmatrix} a_1 \\ a_2 \end{pmatrix}$$

which is particularly expressive, since it tells us that (a_1, a_2) are the coordinates of v with respect to the system whose vectors of unit length form F. This way of writing the coordinates as the column of the column matrix M_v^F allows us to read the formula (1) as

$$v = F \cdot M_v^F \tag{2}$$

What happens if, instead of a vector v we have a pair of vectors $S = (v_1, v_2)$? For each of these vectors formula (2) is valid and hence we have

$$S = F \cdot M_S^F \tag{3}$$

One sees, yet again, the efficacy of using the row by column product. But, the best is yet to come. In fact, we can apply formula (3) to the pair $G = (v_1, v_2)$ and thus obtain

$$G = F \cdot M_G^F \tag{4}$$

What happens if the coordinates of v with respect to G are (b_1, b_2)? We can write, as above, $M_v^G = \binom{b_1}{b_2}$ and $v = G \cdot M_v^G$. If we transform the expression $v = G \cdot M_v^G$ by substituting (4) into it, we obtain

$$v = F \cdot M_G^F M_v^G \tag{5}$$

But, since we also have $v = F \cdot M_v^F$ (see (2)), the uniqueness of the coordinates implies the following equality

$$M_v^F = M_G^F M_v^G \tag{6}$$

We come to the conclusion of this discussion of change of coordinates by putting together the preceding formulas. Thus we have

$$\begin{pmatrix} a_1 \\ a_2 \end{pmatrix} = M_G^F \begin{pmatrix} b_1 \\ b_2 \end{pmatrix} + \begin{pmatrix} c_1 \\ c_2 \end{pmatrix} \tag{7}$$

Inasmuch as the preceding discussion could seem a bit dry, let's immediately look at a specific case.

If P, v_1, v_2 have coordinates, $(1, 2)$, $(-1, 4)$, $(2, -11)$ respectively, with respect to $\Sigma(O; u_1, u_2)$, then

$$M_G^F = \begin{pmatrix} -1 & 2 \\ 4 & -11 \end{pmatrix}$$

and thus

$$M_v^F = \begin{pmatrix} -1 & 2 \\ 4 & -11 \end{pmatrix} M_v^G + \begin{pmatrix} 1 \\ 2 \end{pmatrix}$$

If we let x, y be the coordinates of an arbitrary vector u with respect to $\Sigma(O; u_1, u_2)$ and x', y' the coordinates of u with respect to $\Sigma(P; v_1, v_2)$, then we have the formula

$$\begin{pmatrix} x \\ y \end{pmatrix} = \begin{pmatrix} -1 & 2 \\ 4 & -11 \end{pmatrix} \begin{pmatrix} x' \\ y' \end{pmatrix} + \begin{pmatrix} 1 \\ 2 \end{pmatrix} \tag{8}$$

that is

$$\begin{cases} x = -x' + 2y' + 1 \\ y = 4x' - 11y' + 2 \end{cases} \tag{9}$$

This chapter concludes with the following brief, but intense, section. The reader will have to concentrate in order to follow the mathematical reasoning which allows us to insert all that we have just seen about changes of coordinates into a much broader context.

4.8 Vector Spaces and Bases

This last section of the first part of the book plays a *basic* role, inasmuch as it not only completes the answers to some previous questions, but it also gives the *basis* for many of the discussions in the second part of the book (*translator's note: the puns are in the original and are intended!*).

Let's start off with the following question: how would the discussion of the previous section change if, instead of the plane, we considered the same problem in $\mathbb{R}^n$? But, first of all, we should ask ourselves: what do we mean by a coordinate system in $\mathbb{R}^n$? Let's remember that if a vector v (in the plane) has coordinates (a_1, a_2) in the system of coordinates $\Sigma(O; u_1, u_2)$, then one may write v uniquely as $v = a_1 u_1 + a_2 u_2$. The essential ingredient is the fact that there exist vectors u_1, u_2, in the plane, such that any vector can be written in a unique way as a *constant multiple of the first vector added to a constant multiple of the second vector*. It is worthwhile to give a special name to a sum of the type $a_1 u_1 + a_2 u_2 + \cdots + a_r u_r$, i.e. a sum of vectors $u_1, u_2, \ldots, u_r$ multiplied by constants $a_1, a_2, \ldots, a_r$. One calls such a sum a **linear combination** of the vectors $u_1, u_2, \ldots, u_r$. If we consider the set of n-tuples, that is the set K^n (with K a number field) we see that in it we can do the operations which allow us to construct linear combinations. Sets in which we can form linear combinations, like K^n, are called **vector spaces**. In fact, mathematicians consider many other types of vector spaces, but as far as we are concerned, the spaces of n-tuples are more than sufficient.

The equivalent notion to that of 1-, 2-, or 3-tuple of unit vectors which defined a coordinate system on the line, in the plane and in space is thus, in K^n, an n-tuple of **vectors** $F = (f_1, f_2, \ldots, f_n)$, with the property that every vector in K^n can be written in a unique way as a linear combination of the vectors of F. Such an n-tuple of vectors is called a **basis** of K^n. Here we are at the fundamental point where geometry can no longer help with its pictures, as it did on the line and in the plane. It is precisely at this point where algebra comes in to generalize the concept of a system of coordinates using the concept of a basis.

But, do bases really exist? Have no fear, they exist in abundance. In fact, there is one that is so obvious that it even has a name, we call it the **canonical basis**. This is the basis $E = (e_1, e_2, \ldots, e_n)$ where the vectors $e_1, \ldots, e_n$ are defined by

$$e_1 = (1, 0, \ldots, 0), \; e_2 = (0, 1, 0, \ldots, 0), \ldots, \; e_n = (0, \ldots, 0, 1)$$

One notices how easy it is to find the coordinates of a vector with respect to E. In fact, if $u = (a_1, a_2, \ldots, a_n)$ is a vector of K^n then one has $u = a_1 e_1 + a_2 e_2 + \cdots + a_n e_n$, that is **the coordinates coincide with the components**.

We've said that there are lots of bases. How can we find them and how can we recognize them? Let's suppose that we are given an r-tuple $F = (f_1, f_2, \ldots, f_r)$ of vectors in K^n. As we have already seen in the case of vectors in the plane, we can write their coordinates with respect to the canonical basis in a matrix of size $n \times r$. It seems natural to use the same notation that we introduced in the case of vectors in the plane and thus we have

$$F = E \cdot M_F^E \tag{a}$$

The reasoning now goes as follows: to say that F is a basis means that every vector of K^n can be written in a unique way as a linear combination of the vectors in F. In particular, one can do this for the vectors of E. Thus we can find a matrix, which we naturally call M_E^F, such that

$$E = F \cdot M_E^F \tag{b}$$

If $S = (v_1, v_2, \ldots, v_s)$ is an s-tuple of vectors, we then have

$$S = E \cdot M_S^E \qquad S = F \cdot M_S^F \tag{c}$$

Substituting (a) in the second equation of (c) and substituting (b) in the first equation of (c) we get

$$S = E \cdot M_F^E M_S^F \qquad S = F \cdot M_E^F M_S^E \tag{d}$$

Comparing (c) with (d) and taking into account the uniqueness of the representation with respect to a basis, we get the equality of matrices

$$M_S^E = M_F^E M_S^F \qquad M_S^F = M_E^F M_S^E \tag{e}$$

In particular, one has

$$I_n = M_E^E = M_F^E M_E^F \qquad I_r = M_F^F = M_E^F M_F^E \tag{f}$$

We already mentioned in Section 2.5 that the two relations (f) have as a consequence the fact that the matrices M_F^E and M_E^F are square and inverses of each other, in other words

$$M_E^F = (M_F^E)^{-1} \tag{g}$$

We have found the condition we have been searching for! Using rule (g) of Section 4.6 we can deduce the following property.

(1) **An r-tuple of vectors forms a basis if and only if the matrix associated to it with respect to the canonical basis is invertible. In such an instance one has $r = n$.**

(2) **An n-tuple of vectors forms a basis if and only if the determinant of the matrix associated to it with respect to the canonical basis, is different from zero.**

Once again, the notion of an invertible matrix makes a significant "curtain call". Moreover, from property (1) we deduce the following very important fact.

All the bases of K^n consist of n vectors.

Naturally the converse is not true, in other words, it's not true that every set of n vectors in K^n is a basis. That would be like thinking that every square matrix was invertible and we already know that this is not the case. As we have always done, let's look at some examples with the aim of familiarizing ourselves with the mathematical facts we have just discussed.

Example 4.8.1. Let's consider the following pair of vectors $S = (v_1, v_2)$ in $\mathbb{R}^2$, with $v_1 = (1,1)$, $v_2 = (2,2)$. We have

$$M_S^E = \begin{pmatrix} 1 & 2 \\ 1 & 2 \end{pmatrix}$$

and we notice that M_S^E is not invertible since its determinant is zero. Thus, S is not a basis of $\mathbb{R}^2$. The geometrical explanation for this is found by observing that if $(1,1)$, $(2,2)$ represent the coordinates of two vectors in a plane with cartesian coordinates, then the two vectors are collinear and thus their linear combinations give us only vectors on a line and not all the vectors in the plane.

Example 4.8.2. Let's consider the following triple of vectors $S = (v_1, v_2, v_3)$ in $\mathbb{R}^2$, with $v_1 = (1,2)$, $v_2 = (1,0)$, $v_3 = (0,2)$. One has

$$M_S^E = \begin{pmatrix} 1 & 1 & 0 \\ 2 & 0 & 2 \end{pmatrix}$$

We've already observed that three vectors are too many to constitute a basis for $\mathbb{R}^2$. And, in fact, we see that $v_1 - v_2 - v_3 = 0$ and hence $v_1 - v_2 - v_3 = 0v_1 - 0v_2 - 0v_3$ are two different representations of the zero vector. That contradicts the fundamental property which every basis possesses, i.e. that every vector in $\mathbb{R}^n$ may be written in a unique way as a linear combination of the vectors of S.

Example 4.8.3. Consider the following pair of vectors $S = (v_1, v_2)$ in $\mathbb{R}^3$, with $v_1 = (1, 2, 3)$, $v_2 = (1, 2, 4)$. We have

$$M_S^E = \begin{pmatrix} 1 & 1 \\ 2 & 2 \\ 3 & 4 \end{pmatrix}$$

and we see that M_S^E is not invertible, since it is not square. In this case there are too few vectors in S and it would be too much to believe that linear combinations of just two vectors could give us everything in $\mathbb{R}^3$. The geometric picture to keep in mind is the following: linear combinations of two non-parallel vectors in spaces will give us all the vectors of a plane and hence cannot fill up all of space.

An amplification of the ideas considered here comes up in Part II in Section 6.3, but we can immediately extract some benefits from the ideas just mentioned. Let E be the canonical basis of K^n, and let $F = (f_1, \ldots, f_n)$ be any basis of K^n and $v = (a_1, \ldots, a_n)$ a vector of K^n. We already know that v may be written in a unique way as a linear combination of the vectors in E and so $M_v^E = (a_1 \cdots a_n)^{\mathrm{tr}}$. Moreover, we can write the vector v in a unique fashion as a linear combination of the vectors in F, in other words, there are uniquely determined numbers $b_1, \ldots, b_n \in K$ such that $v = b_1 f_1 + \cdots + b_n f_n$. Put another way, there are uniquely determined numbers $b_1, \ldots, b_n \in K$ such that $M_v^F = (b_1 \cdots b_n)^{\mathrm{tr}}$. Thus we have the following identity

$$v = E \cdot M_v^E \qquad v = F \cdot M_v^F$$

Now we are in a position to answer the following question: what relationship is there between the coordinates of v with respect to E and the coordinates of v with respect to F? This question is a natural extension of the question we posed in Section 4.7 about changes of coordinates. The answer is rather simple. In fact,

$$F \cdot M_v^F = E \cdot M_v^E = F \cdot M_E^F M_v^E = F \cdot (M_F^E)^{-1} M_v^E$$

where the second inequality follows from formula (b) and the third from formula (g) that we saw above. Once more we invoke the uniqueness of the representation and deduce the following formula which computationally answers our question

$$M_v^F = (M_F^E)^{-1} M_v^E \tag{h}$$

We conclude this section, and thus both the chapter and Part I, with an example that highlights the importance of formula (h).

Example 4.8.4. Consider $v = (-2, 1, 6)$, $v_1 = (1, 1, 1)$, $v_2 = (0, 1, 0)$, $v_3 = (1, 0, 2)$, vectors in $\mathbb{R}^3$, and let $F = (v_1, v_2, v_3)$. We have

$$M_F^E = \begin{pmatrix} 1 & 0 & 1 \\ 1 & 1 & 0 \\ 1 & 0 & 2 \end{pmatrix}$$

Formula (h) says that $M_v^F = (M_F^E)^{-1} M_v^E$. We need only calculate the matrix $(M_F^E)^{-1}$, which is

$$(M_F^E)^{-1} = \begin{pmatrix} 2 & 0 & -1 \\ -2 & 1 & 1 \\ -1 & 0 & 1 \end{pmatrix}$$

Thus, we have

$$M_v^F = \begin{pmatrix} 2 & 0 & -1 \\ -2 & 1 & 1 \\ -1 & 0 & 1 \end{pmatrix} \begin{pmatrix} -2 \\ 1 \\ 6 \end{pmatrix} = \begin{pmatrix} -10 \\ 11 \\ 8 \end{pmatrix}$$

By the way, it's easy to check that $v = -10v_1 + 11v_2 + 8v_3$.

For the moment let's be content with the considerations and the examples we have already seen. As I mentioned above, we'll come back to them and develop them in more depth in Section 6.3. Some readers might object to this way of doing things, indeed it is not linear. But, as has already been observed, the world in which we live is not linear and neither are...books of linear algebra.

Exercises

Exercise 1. Consider the vectors $v_1 = (1,0,0)$, $v_2 = (-1,-1,2)$, $v_3 = (0,0,1)$ of $\mathbb{R}^3$ and the triple $F = (v_1, v_2, v_3)$. Let

$$A = \begin{pmatrix} 1 & 1 & -2 \\ 2 & -2 & 1 \\ 1 & 0 & 0 \end{pmatrix}$$

(a) Prove that F is a basis of $\mathbb{R}^3$.
(b) Knowing that $A = M_F^G$, determine G.

Exercise 2. Let's suppose that we are given a monometric orthogonal coordinate system in the plane and consider the vectors $u = (2,2)$, $v = (-1,-2)$.

(a) Calculate the cosine of the angle formed by u and normal(v).
(b) Find three vectors which have the same modulus as v.
(c) Describe the set of all vectors which are perpendicular to u.

Exercise 3. Consider the vectors $v_1 = (1,2,0)$, $v_2 = (1,-1,1)$, $v_3 = (0,1,2)$, $u = (1,1,1)$ of $\mathbb{R}^3$ and the triple $F = (v_1, v_2, v_3)$.

(a) Verify that F is a basis of $\mathbb{R}^3$.
(b) Calculate the volume of the parallelepiped defined by v_1, v_2, v_3.
(c) Verify the equality $u = E \cdot M_F^E M_u^F$.

Exercise 4. Consider the following matrices

$$A = \begin{pmatrix} 1 & 1 & 0 \\ 3 & -1 & -2 \\ 0 & 0 & \frac{1}{2} \end{pmatrix}$$

Does there exist a basis F of $\mathbb{R}^3$ such that $A = M_E^F$?

Exercise 5. Suppose that we have a monometric orthogonal coordinate system in space and the vectors $u_1 = (1,2,0)$, $u_2 = (2,4,1)$, $u_3 = (4,9,1)$ of $\mathbb{R}^3$.

(a) Calculate the volume of the parallelepiped defined by the three vectors.
(b) Find all the vectors which are perpendicular to both u_1 and u_2.
(c) Find a vector v with the following two properties:

$$v = u_2 \qquad u_1 \cdot u_2 < u_1 \cdot v$$

Exercise 6. Using the properties of the determinant, prove that if A is an invertible matrix the following formula is true: $\det(A^{-1}) = (\det(A))^{-1}$.

Exercise 7. Let n be a positive whole number.

(a) Let $u, v \in \mathbb{R}^n$ be two vectors with rational components. Is it true that their scalar product is a rational number?

(b) Is it true that if a non-zero vector in $\mathbb{R}^n$ has rational components then the unit vector in the same direction has rational components?

Exercise 8. Consider the two matrices

$$A = \begin{pmatrix} 1 & 3 & \frac{1}{2} \end{pmatrix}^{\mathrm{tr}} \qquad B = \begin{pmatrix} 1 & -1 & 4 \end{pmatrix}$$

Verify that $\det(A \cdot B) = 0$ and find a geometric explanation for that.

Exercise 9. Suppose that we are given a monometric orthogonal coordinate system in space and the five points $A_1 = (3, 0, 0)$, $A_2 = (1, 3, 3)$, $A_3 = (0, 0, 2)$, $A_4 = (0, 7, 0)$, $A_5 = (1, 1, 1)$.

(a) What are the two closest points of these five?

(b) If the distance formula between two points in space were $|b_1 - a_1| + |b_2 - a_2| + |b_3 - a_3|$, instead of $\sqrt{(b_1 - a_1)^2 + (b_2 - a_2)^2 + (b_3 - a_3)^2}$, would you have the same response to the previous question?

@ **Exercise 10.** Let A be a 5×5 matrix with random entries. (To make such a matrix you should use a computer algebra system, for example CoCoA.) Using CoCoA you can form the matrix M as follows:

```
L:=[[Randomized(X) | X In 1..5] | X In 1..5]; M:=Mat(L);
```

(a) Calculate the determinant of A and observe that it is not equal to zero.

(b) Repeat the experiment, i.e. construct another random matrix, and observe that the same thing happens, i.e. its determinant is not equal to zero.

(c) Give an explanation of the fact that a square matrix with random entries is invertible.

Observation: *If the reader finds a matrix (using the procedure above) whose determinant is equal to zero, the possible reasons for this are two: the reader has made a mistake interfacing with his calculator or the reader is gifted with paranormal powers.*

* *

This finishes Part I. If the reader is, by chance, thinking of stopping here, I would like to try and dissuade him or her. What follows is certainly a little more difficult to read but I really believe it is worth the effort. The strength of linear algebra, which up to now has been contained, will begin to reveal itself more completely.

I parte, II arriva

(piccola anticipazione della **Parte II**)

Part II

5

Quadratic Forms

this is □n ex□mple of □ qu□dr□tic f□rm

In this chapter we will study yet another aspect of the extraordinary capacity of matrices to adapt themselves to very diverse situations.

In the same way that systems of linear equations can be used as mathematical models for a multitude of problems, polynomial systems i.e. a system that we obtain by setting equal to zero a finite number of multivariate polynomials, are even more important and permit us to model a still larger number of problems.

Just to give a simple (self-referential) example of this, it turns out that the design of the fonts in which this book is written is done using curves which can be described analytically by polynomial equations of degree less than or equal to three. We are not going to go into that here, but I wanted to mention it just to give an indication of the importance of describing, with equations, entities that will then appear on a screen or on paper as graphic objects.

As another example, if we wanted to describe a circle with center at the point $P(1, 2)$ and with radius 3 using a monometric system of coordinates in the plane, we could use the equation

$$(x - 1)^2 + (y - 2)^2 - 9 = 0$$

It's clear that drawing that circle on a monitor would require the coloring of a large number of pixels. But, how do you explain to a printer or to a collaborator by means of e-mail the information about how to draw that circle? A simple way would be to transmit a list of all the pixels that you had to color.

For example, if the page had *length* 6000 pixels and *width* 4000 pixels and we only wanted to transmit the picture in black and white, it would be enough to transmit a matrix A of type 6000×4000 whose white pixels could be coded by putting a zero in the appropriate place in the matrix and by putting a 1 in the matrix where we wanted a black pixel. This method

Robbiano L.: Linear Algebra for everyone
© Springer-Verlag Italia 2011

would be ok no matter what the figure was. Naturally, the information about the black circumference on a white background could be made much more compact using more subtle methods. How?

One way is to transmit the three numbers $1, 2, 3$. The person who receives these numbers knows that the first two numbers represent the center of the circle and the third number represents the radius! Instead of the millions of entries in the matrix A, it was enough to have three numbers! But, things don't finish here. Suppose we wanted to magnify our illustration? Using the mathematical description *it is sufficient just to change one parameter*, namely the radius of the circle.

This example, notwithstanding the fact that it is strongly simplified, should give an idea of how mathematics can be an essential support for technology. Unfortunately, the treatment of the mathematical objects described by systems of polynomial equations is much more complicated that those described by linear systems. But, there is one very important case where linear algebra with its toolkit of ideas, in particular with the powerful idea of a matrix, returns to center stage. We'll see this in the study of equations of the second degree.

5.1 Equations of the Second Degree

I guess that *all* the readers know what an equation of the second degree is, or at least they believe that they know what it is. Let's see if that's true by, for instance, trying to represent a generic equation of the second degree in three variables.

The answer is $f(x, y, z) = 0$ with

$$f(x, y, z) = a_1 x^2 + a_2 y^2 + a_3 z^2 + a_4 xy + a_5 xz + a_6 yz + a_7 x + a_8 y + a_9 z + a_{10}$$

Let's look carefully at the polynomial $f(x, y, z)$ and begin by saying that its *leading* part is the homogeneous part of the second degree, i.e.

$$a_1 x^2 + a_2 y^2 + a_3 z^2 + a_4 xy + a_5 xz + a_6 yz$$

In what sense is it *leading*? Doing some reasoning similar to what we did in Section 3.4 we can think of the polynomial as a function and then the linear part

$$a_7 x + a_8 y + a_9 z + a_{10}$$

is overcome by the quadratic part when x, y and z get big.

Another observation is that, if we add a new variable w, we can *homogenize* the polynomial and write

$$a_1 x^2 + a_2 y^2 + a_3 z^2 + a_4 xy + a_5 xz + a_6 yz + a_7 xw + a_8 yw + a_9 zw + a_{10} w^2$$

To get back the original polynomial all we have to do is make the substitution $w = 1$. We won't dwell now on the mathematical subtleties that arise in these considerations. Suffice it to say that in order to analyze equations of the second degree, the mathematical objects which are most important to study are the homogeneous polynomials of the second degree, also known as **quadratic forms**.

What, then, is a quadratic form? Mathematicians would define them correctly saying that a quadratic form is a *sum of monomials of the second degree*. In the case of two variables, x and y, a quadratic form is thus an expression of the type

$$ax^2 + bxy + cy^2$$

And why are these so important? And, moreover, what do they have to do with linear algebra? Let's start by observing that we can also speak of **linear forms**. A linear form, inasmuch as it is a sum of monomials of degree 1, is nothing more than a polynomial of degree one with constant term equal to zero, thus it is an expression of the type

$$\alpha x + \beta y$$

which we have already encountered and studied in the first part of the book.

Now let's make a small observation. If $\alpha x + \beta y$ is a linear form, then its square is $(\alpha x + \beta y)^2 = \alpha^2 x^2 + 2\alpha\beta xy + \beta^2 y^2$ and thus is a quadratic form. If, at this moment, you might think that all quadratic forms are nothing more than squares of linear forms, it would be best to not have that thought again! Anyone who has studied a small bit of geometry knows that this is not true because there are conics which are not double lines, in fact the *overwhelming number* of conics are not double lines. For those of you who have not studied any geometry, the preceding remark will not mean much. In that case you will have to depend on some ideas from high school where you probably saw that the quadratic form $x^2 + xy + y^2$ was not the square of a linear form.

Nevertheless, the question still remains open: how are quadratic forms related to linear algebra? The relationship comes from the following equation which mathematicians noticed

$$ax^2 + bxy + cy^2 = (x \quad y) \begin{pmatrix} a & \frac{b}{2} \\ \frac{b}{2} & c \end{pmatrix} \begin{pmatrix} x \\ y \end{pmatrix}$$

which is, at first glance, unexpected and esthetically even ugly. If we call $2b$ (instead of b) the coefficient of xy the relation can be rewritten as

$$ax^2 + 2bxy + cy^2 = (x \quad y) \begin{pmatrix} a & b \\ b & c \end{pmatrix} \begin{pmatrix} x \\ y \end{pmatrix} \tag{$*$}$$

which is at least a bit prettier to look at. Some readers might wonder if we have saved some of the beauty of the matrix notation at the expense of limiting the validity of the formula. Perhaps the writer of the second formula thinks that every number is a multiple of two? But we know, for example, that in $\mathbb{Z}$ the number 3 is not a multiple of 2. The fact is that every number can be written as the double of its half, *assuming, of course, that the half exists!* We will guarantee that *one half* exists if we consider our coefficients as being in a number field whose characteristic is different from 2 (which guarantees the existence of $\frac{1}{2}$).

The reader shouldn't worry over much if the preceding phrases are not well understood. It will be enough to understand that, for example, $\mathbb{Q}$ and $\mathbb{R}$ are fine for us and we may, without worry, put aside this mathematical subtlety.

With this observation in mind and in order to avoid difficulties (and for other reasons which we will discuss later) we shall, *in this chapter, always take our coefficients from the field $\mathbb{R}$ of real numbers.*

We finally come to the expression $(*)$ above. Given that x and y are the names given to the unknowns, all the information of the quadratic form is contained in the matrix $A = \begin{pmatrix} a & b \\ b & c \end{pmatrix}$, which we immediately notice is *symmetric* (see Section 2.2). Here, once again, we are coming face to face with matrices and we can see the analogy with what was said about the matrix associated to a linear system. Just as in the earlier situation, you'll recall, all the information is recorded in one particular matrix. The preceding observation made for quadratic forms in two variables can be generalized. If we have n variables $x_1, x_2, \ldots, x_n$ and write the generic quadratic form

$$Q = a_{11}x_1^2 + 2a_{12}x_1x_2 + \cdots + a_{nn}x_n^2$$

then we again have the equality

$$Q = \begin{pmatrix} x_1 & x_2 & \cdots & x_n \end{pmatrix} \begin{pmatrix} a_{11} & a_{12} & \cdots a_{1n} \\ a_{12} & a_{22} & \cdots a_{2n} \\ \cdots & \cdots & \cdots \\ a_{1n} & a_{2n} & \cdots a_{nn} \end{pmatrix} \begin{pmatrix} x_1 \\ x_2 \\ \vdots \\ x_n \end{pmatrix} \tag{1}$$

The symmetric matrix obtained in this way is called **the matrix of the quadratic form**. Let's look at an example.

Example 5.1.1. The quadratic form

$$Q = 2x^2 - \frac{1}{3}xy + 2yz - z^2$$

may be written

$$Q = \begin{pmatrix} x & y & z \end{pmatrix} \begin{pmatrix} 2 & -\frac{1}{6} & 0 \\ -\frac{1}{6} & 0 & 1 \\ 0 & 1 & -1 \end{pmatrix} \begin{pmatrix} x \\ y \\ z \end{pmatrix}$$

as can be verified directly by carrying out the various products.

Now, let's prepare ourselves for another change of scene! If $(x_1, x_2, \ldots, x_n)$ is thought of as a general vector then we know that its components are also the coordinates with respect to the canonical basis.

This observation can be viewed as thinking of the expression (1) as making reference to the canonical basis $E = (e_1, e_2, \ldots, e_n)$. That, in turn, suggests that we call the symmetric matrix above M_Q^E. Why M_Q^E? Some would say that this notation is rather cumbersome, which in fact it is, but (as often happens) one pays a high price for a clear expression. In this case we are paying (with the awkwardness of the notation) for the fact that it expresses the ideas clearly. What I mean is that M_Q^E self describes the fact that it is the *matrix of the quadratic form Q with reference to the canonical basis E*.

How does such a description help us? Writing $v = (x_1, x_2, \ldots, x_n)$, the quadratic form can then be written in the following way

$$Q = (M_v^E)^{\text{tr}} \, M_Q^E \, M_v^E \tag{2}$$

And exactly where is the advantage of interpreting the quadratic form in this way which, at first glance, seems more abstruse? But, suppose we consider another basis $F = (v_1, v_2, \ldots, v_n)$ of $\mathbb{R}^n$. We have seen in Section 4.8 that the fact of being a basis can be translated into saying that the matrix M_F^E is invertible. We've also seen that $M_v^E = M_F^E \, M_v^F$. So, if we substitute these expressions into the equality (2), we obtain

$$Q = (M_F^E \, M_v^F)^{\text{tr}} \, M_Q^E \, M_F^E M_v^F = (M_v^F)^{\text{tr}} \, (M_F^E)^{\text{tr}} \, M_Q^E \, M_F^E \, M_v^F \tag{3}$$

And now we have a nice surprise. The preceding formula expresses the fact that if the matrix of the quadratic form Q with respect to E is M_Q^E, then the matrix of the same quadratic form with respect to the basis F is $(M_F^E)^{\text{tr}} \, M_Q^E \, M_F^E$. We have thus proved the following formula

$$M_Q^F = (M_F^E)^{\text{tr}} \, M_Q^E \, M_F^E \tag{4}$$

At this stage it's not very clear how one could possibly use such a formula and, before going ahead with our study, let's at least see if we can understand the underlying idea. For the moment, let's be content with knowing that having a formula like (4) at our disposal, we could hope to find an opportune basis F so that the matrix $(M_F^E)^{\text{tr}} \, M_Q^E \, M_F^E$ is *easier or nicer* than the matrix M_Q^E. Studying the matrices associated to a linear system, we have already learned that *easier or nicer* means *lots of zeroes* and, in fact, if the matrix $(M_F^E)^{\text{tr}} \, M_Q^E \, M_F^E$ has lots of zeroes, the quadratic form also has lots of zero coefficients and is thus easier to describe.

This discussion has, necessarily, been a bit vague and so we should really look at an example.

Example 5.1.2. Making the mixed term disappear
Let's consider the quadratic form $Q = 3x^2 - 4xy + 3y^2$ in the variables x, y. Let $v_1 = (\frac{\sqrt{2}}{2}, \frac{\sqrt{2}}{2})$, $v_2 = (-\frac{\sqrt{2}}{2}, \frac{\sqrt{2}}{2})$ and let $F = (v_1, v_2)$. One has the equality $M_F^E = \begin{pmatrix} \frac{\sqrt{2}}{2} & -\frac{\sqrt{2}}{2} \\ \frac{\sqrt{2}}{2} & \frac{\sqrt{2}}{2} \end{pmatrix}$ and since $\det(M_F^E) = 1$ we deduce that the matrix

is invertible and hence F is a basis of $\mathbb{R}^2$. What happens if we write the quadratic form Q with respect to the basis F? Formula (4) gives the answer and hence one has

$$M_Q^F = \begin{pmatrix} \frac{\sqrt{2}}{2} & \frac{\sqrt{2}}{2} \\ -\frac{\sqrt{2}}{2} & \frac{\sqrt{2}}{2} \end{pmatrix} \begin{pmatrix} 3 & -2 \\ -2 & 3 \end{pmatrix} \begin{pmatrix} \frac{\sqrt{2}}{2} & -\frac{\sqrt{2}}{2} \\ \frac{\sqrt{2}}{2} & \frac{\sqrt{2}}{2} \end{pmatrix} = \begin{pmatrix} 1 & 0 \\ 0 & 5 \end{pmatrix}$$

This means that, if the calculations have been done well and the new coordinates of the vector v are called x', y' the quadratic form can be written $x'^2 + 5y'^2$. Let's check that.

We have

$$\begin{pmatrix} x \\ y \end{pmatrix} = M_v^E = M_F^E M_v^F = \begin{pmatrix} \frac{\sqrt{2}}{2} & -\frac{\sqrt{2}}{2} \\ \frac{\sqrt{2}}{2} & \frac{\sqrt{2}}{2} \end{pmatrix} \begin{pmatrix} x' \\ y' \end{pmatrix}$$

Consequently, we get

$$x = \frac{\sqrt{2}}{2}(x' - y') \qquad y = \frac{\sqrt{2}}{2}(x' + y')$$

Now let's substitute this in the expression $Q = 3x^2 - 4xy + 3y^2$ to obtain

$$Q = 3(\frac{\sqrt{2}}{2}(x'-y'))^2 - 4(\frac{\sqrt{2}}{2}(x'-y'))(\frac{\sqrt{2}}{2}(x'+y')) + 3(\frac{\sqrt{2}}{2}(x'+y'))^2 = x'^2 + 5y'^2$$

Notice that the coefficient of $x'y'$ is zero and hence *the mixed term has disappeared.* The wise use of matrices has allowed us to find a new system of coordinates so that with respect to that new system the quadratic form is simpler. More precisely, in the first expression we had the mixed term xy and now the mixed term $x'y'$ doesn't appear.

This example, interesting though it seems, is not yet satisfying. In fact, *all the readers* will have noticed that the choice of the new basis seemed to come out of nowhere. What oracle has told us to choose the vectors $v_1 = (\frac{\sqrt{2}}{2}, \frac{\sqrt{2}}{2})$, $v_2 = (-\frac{\sqrt{2}}{2}, \frac{\sqrt{2}}{2})$?

In order to understand how we chose that basis the reader has to be patient, as we have quite a long road to travel before we see how that was done. Certainly one doesn't choose these bases by trial and error, we need to have a method. Nevertheless, before concluding this section it is useful to make one more observation which is contained in the following example.

Example 5.1.3. The Identity Matrix

What happens if $M_Q^E = I_n$? We can give the answer quickly and easily. In fact, if $v = (x_1, x_2, \ldots, x_n)$ is the generic vector we have the following equality

$$Q = \begin{pmatrix} x_1 & x_2 & \cdots & x_n \end{pmatrix} I_n \begin{pmatrix} x_1 \\ x_2 \\ \vdots \\ x_n \end{pmatrix} = x_1^2 + x_2^2 + \cdots + x_n^2 = |v|^2$$

Thus, we have easily discovered that if the matrix of the quadratic form is the identity, then the form itself is nothing other than the one we used to define the length of a vector and hence is related to the notion of distance. One more time we have seen that geometry and algebra are intimately tied together and each illuminates the other.

5.2 Elementary Operations on Symmetric Matrices

The preceding section brought to light a fundamental connection between quadratic forms and symmetric matrices. It's time now to go into that link more deeply.

So, let A be a real symmetric matrix of type n. We have seen that A may be thought of as the matrix of a quadratic form in n variables $x_1, \ldots, x_n$, with respect to the canonical basis $E = (e_1, \ldots, e_n)$ of $\mathbb{R}^n$. We have also noted that a change of basis *modifies* A, transforming it into the matrix $P^{\mathrm{tr}} A P$, where $P = M_F^E$. Assuming our objective is that of *simplifying* A, we have to figure out how to use the operation $P^{\mathrm{tr}} A P$ with P invertible, in a way that will achieve our objective. Keep in mind what we did with Gaussian reduction. Recall, that procedure allowed us to manipulate a matrix A into upper triangular form using elementary row operations. By putting together such operations, i.e. by multiplying by the corresponding elementary matrices, we were able to obtain $PA = U$ where P is invertible (because it was a product of elementary matrices) and U is upper triangular.

Can we do the same thing in this new situation? Certainly not, because doing such a thing would destroy the symmetry of the matrix A. We have to figure out a new strategy which will bring us **not** to PA but to $P^{\mathrm{tr}} A P$.

The key observation is that if we do an elementary operation on the rows and *the same elementary operation on the columns*, we don't destroy the symmetry. In fact, if an elementary operation on the rows has a certain effect, the *symmetric* effect can be obtained by doing the corresponding elementary operation on the columns. How can we be sure that this is correct? Mathematicians usually try to give a rigorous proof, but first we have to understand what it is we want to prove.

So, observe that if M and N are two matrices for which we can form the product MN, then

$$(MN)^{\mathrm{tr}} = N^{\mathrm{tr}} M^{\mathrm{tr}} \tag{1}$$

Also observe that for every matrix A we have the following equality

$$(A^{\mathrm{tr}})^{\mathrm{tr}} = A \tag{2}$$

The following fact is also true

$$A \text{ is symmetric if and only if } A = A^{\mathrm{tr}} \tag{3}$$

Finally notice that if A is invertible then we have the equality

$$(A^{\text{tr}})^{-1} = (A^{-1})^{\text{tr}} \tag{4}$$

The proofs of these last statements are not going to be given here, but we suggest that the reader try to give the proofs because all are actually rather easy.

Now, putting all these various facts together we get: if A is a symmetric matrix and B is a square matrix of the same type as A then the matrix $B^{\text{tr}} A B$ is symmetric. In fact it is enough to use rules (1) and (2) to obtain the equality

$$(B^{\text{tr}} A B)^{\text{tr}} = B^{\text{tr}} A^{\text{tr}} (B^{\text{tr}})^{\text{tr}} = B^{\text{tr}} A B$$

from which the conclusion follows by applying rule (3).

Now that we know that performing operations of the type $B^{\text{tr}} A B$ doesn't destroy the symmetry of A, let's return to the strategy of doing elementary operations on the rows and columns. We'll start trying to understand what happens by working out some examples.

Example 5.2.1. Let $A = \left(\begin{smallmatrix} 1 & 2 \\ 2 & 3 \end{smallmatrix}\right)$. We can use $a_{11} = 1$ as a pivot and take our first step toward Gaussian reduction by subtracting from the second row twice the first row. We know that this means multiplying, on the left, by the elementary matrix $E_1 = \left(\begin{smallmatrix} 1 & 0 \\ -2 & 1 \end{smallmatrix}\right)$. One obtains $B = E_1 A = \left(\begin{smallmatrix} 1 & 2 \\ 0 & -1 \end{smallmatrix}\right)$, which is no longer symmetric. So, let's consider $(E_1)^{\text{tr}} = \left(\begin{smallmatrix} 1 & -2 \\ 0 & 1 \end{smallmatrix}\right)$ and multiply B on the right by it, obtaining $B(E_1)^{\text{tr}} = \left(\begin{smallmatrix} 1 & 0 \\ 0 & -1 \end{smallmatrix}\right)$. We've ended up with a matrix that is symmetric and has lots of zeroes. In conclusion, putting $P = (E_1)^{\text{tr}}$ one has

$$P = \begin{pmatrix} 1 & -2 \\ 0 & 1 \end{pmatrix} \qquad P^{\text{tr}} A P = \begin{pmatrix} 1 & 0 \\ 0 & -1 \end{pmatrix}$$

If we think of A as the matrix of the quadratic form $x^2 + 4xy + 3y^2$ then, with respect to the new basis $F = (v_1, v_2)$, where $v_1 = e_1$, $v_2 = (-2, 1)$, the same form can now be written $x'^2 - y'^2$.

Having understood this last example, we are at a good point. Now we know that we can proceed with the Method of Gauss as long as we can find a non-zero pivot on the principal diagonal. Obviously, this is not enough. It could happen that in the right place on the principal diagonal there is a zero! What would we do then? In that case, the Gauss method tells us to exchange two appropriate rows. Can we do the same thing in our situation? From what we said earlier, any exchange of rows has to be accompanied by a corresponding exchange of columns. Let's see how that would work.

Example 5.2.2. Let $A = \left(\begin{smallmatrix} 0 & 1 \\ 1 & 2 \end{smallmatrix}\right)$. Inasmuch as $a_{11} = 0$ let's try to exchange the row and also the column. In order to do that put $E_1 = \left(\begin{smallmatrix} 0 & 1 \\ 1 & 0 \end{smallmatrix}\right)$ and so we get

$$(E_1)^{\text{tr}} A E_1 = \begin{pmatrix} 2 & 1 \\ 1 & 0 \end{pmatrix}$$

This matrix has a non-zero pivot in the right place and thus we can proceed as in the previous example. One obtains

$$\begin{pmatrix} 1 & 0 \\ -\frac{1}{2} & 1 \end{pmatrix} \begin{pmatrix} 2 & 1 \\ 1 & 0 \end{pmatrix} \begin{pmatrix} 1 & -\frac{1}{2} \\ 0 & 1 \end{pmatrix} = \begin{pmatrix} 2 & 1 \\ 0 & -\frac{1}{2} \end{pmatrix} \begin{pmatrix} 1 & -\frac{1}{2} \\ 0 & 1 \end{pmatrix} = \begin{pmatrix} 2 & 0 \\ 0 & -\frac{1}{2} \end{pmatrix}$$

Thus, putting $P^{\mathrm{tr}} = \begin{pmatrix} 1 & 0 \\ -\frac{1}{2} & 1 \end{pmatrix} \begin{pmatrix} 0 & 1 \\ 1 & 0 \end{pmatrix} = \begin{pmatrix} 0 & 1 \\ 1 & -\frac{1}{2} \end{pmatrix}$, one obtains

$$P^{\mathrm{tr}} A P = \begin{pmatrix} 2 & 0 \\ 0 & -\frac{1}{2} \end{pmatrix}$$

Does that finish everything? No, there is still one more situation in which the previous procedure would not work. Let's take a look at it.

Example 5.2.3. Let $A = \begin{pmatrix} 0 & a \\ a & 0 \end{pmatrix}$ with $a \neq 0$. One can easily see that making an exchange of both rows and columns, nothing changes. In fact

$$\begin{pmatrix} 0 & 1 \\ 1 & 0 \end{pmatrix} \begin{pmatrix} 0 & a \\ a & 0 \end{pmatrix} \begin{pmatrix} 0 & 1 \\ 1 & 0 \end{pmatrix} = \begin{pmatrix} 0 & a \\ a & 0 \end{pmatrix}$$

However, there is a way out of this situation. If we consider the matrix $P = \begin{pmatrix} 1 & -1 \\ 1 & 1 \end{pmatrix}$, we obtain

$$P^{\mathrm{tr}} A P = \begin{pmatrix} 1 & 1 \\ -1 & 1 \end{pmatrix} \begin{pmatrix} 0 & a \\ a & 0 \end{pmatrix} \begin{pmatrix} 1 & -1 \\ 1 & 1 \end{pmatrix} = \begin{pmatrix} 2a & 0 \\ 0 & -2a \end{pmatrix}$$

We have now gathered together all the various instruments we need, and so let's get to work. Let's try to deal with an example which is a bit more difficult.

Example 5.2.4. Let's consider the following real symmetric matrix.

$$A = \begin{pmatrix} 2 & -8 & -3 & -3 \\ -8 & 29 & \frac{79}{6} & \frac{71}{6} \\ -3 & \frac{79}{6} & \frac{437}{108} & \frac{601}{108} \\ -3 & \frac{71}{6} & \frac{601}{108} & \frac{485}{108} \end{pmatrix}$$

We use $a_{11} = 2$ as a pivot to get a zero in the first row, first column. We have

$$(E_1)^{\mathrm{tr}} = \begin{pmatrix} 1 & 0 & 0 & 0 \\ 4 & 1 & 0 & 0 \\ \frac{3}{2} & 0 & 1 & 0 \\ \frac{3}{2} & 0 & 0 & 1 \end{pmatrix} \qquad (E_1)^{\mathrm{tr}} A E_1 = \begin{pmatrix} 2 & 0 & 0 & 0 \\ 0 & -3 & \frac{7}{6} & -\frac{1}{6} \\ 0 & \frac{7}{6} & -\frac{49}{108} & \frac{115}{108} \\ 0 & -\frac{1}{6} & \frac{115}{108} & -\frac{1}{108} \end{pmatrix}$$

Now let's use the entry -3 in the $(2,2)$ position as a pivot. We get

$$(E_2)^{\mathrm{tr}} = \begin{pmatrix} 1 & 0 & 0 & 0 \\ 0 & 1 & 0 & 0 \\ 0 & \frac{7}{18} & 1 & 0 \\ 0 & -\frac{1}{18} & 0 & 1 \end{pmatrix} \qquad (E_2)^{\mathrm{tr}} (E_1)^{\mathrm{tr}} A E_1 E_2 = \begin{pmatrix} 2 & 0 & 0 & 0 \\ 0 & -3 & 0 & 0 \\ 0 & 0 & 0 & 1 \\ 0 & 0 & 1 & 0 \end{pmatrix}$$

This last matrix we can deal with using the method we saw in the preceding example. Thus, one has

$$(E_3)^{\text{tr}} = \begin{pmatrix} 1 & 0 & 0 & 0 \\ 0 & 1 & 0 & 0 \\ 0 & 0 & 1 & 1 \\ 0 & 0 & -1 & 1 \end{pmatrix} \qquad (E_3)^{\text{tr}}(E_2)^{\text{tr}}(E_1)^{\text{tr}} A\, E_1 E_2 E_3 = \begin{pmatrix} 2 & 0 & 0 & 0 \\ 0 & -3 & 0 & 0 \\ 0 & 0 & 2 & 0 \\ 0 & 0 & 0 & -2 \end{pmatrix}$$

Now let's put

$$P = E_1 E_2 E_3 = \begin{pmatrix} 1 & 4 & \frac{13}{3} & -\frac{16}{9} \\ 0 & 1 & \frac{1}{3} & -\frac{4}{9} \\ 0 & 0 & 1 & -1 \\ 0 & 0 & 1 & 1 \end{pmatrix}$$

and we finally have

$$P^{\text{tr}} A\, P = \begin{pmatrix} 2 & 0 & 0 & 0 \\ 0 & -3 & 0 & 0 \\ 0 & 0 & 2 & 0 \\ 0 & 0 & 0 & -2 \end{pmatrix}$$

The conclusion to all of the reasoning above is the following. *Consider the elementary operations on the rows and the corresponding elementary operations on the columns of a real symmetric matrix. To these add a type of elementary operation based on Example 5.2.3. Using these elementary operations, the matrix of any quadratic form may be transformed into a diagonal matrix.* In the language of quadratic forms we can make the following affirmation.

Every real quadratic form may be represented by means of a diagonal matrix.

If the matrix of the quadratic form is A and the elementary matrices which perform the operations on the columns are $E_1, E_2, \ldots, E_r$ then the matrix

$$B = (E_r)^{\text{tr}} \cdots (E_2)^{\text{tr}} (E_1)^{\text{tr}} A\, E_1 E_2 \cdots E_r$$

is diagonal. Putting $P = E_1 E_2 \cdots E_r$ one has

$$B = P^{\text{tr}} A\, P$$

with P invertible and B diagonal.

At this point it is a good idea to know the standard terminology for what we have done above. One says that two real symmetric matrices A and B (of the same type) are **congruent** if there is an invertible matrix P with $B = P^{\text{tr}} A\, P$. In this case we say the two matrices are related by *congruence*. Mathematicians love to observe that this relation is, in fact, an equivalence relation. The preceding affirmation, that every real quadratic form may be represented by a diagonal matrix, is expressed purely in the language of matrices as in the following proposition.

Every real symmetric matrix is congruent to a real diagonal matrix.

With a tiny bit of extra effort we can arrive at a very important observation. We have just seen that if Q is a real quadratic form in n variables then there is a basis F of $\mathbb{R}^n$ such that M_Q^F is diagonal. With the right exchange of both rows and columns we can easily suppose that on this diagonal the first entries are positive numbers, then negative numbers and then zeroes. Is this not completely clear to some readers? Let's immediately look at an example to try to make it clear.

Example 5.2.5. Let's consider the following symmetric matrix

$$A = \begin{pmatrix} 0 & -2 & 1 \\ -2 & -4 & 0 \\ 1 & 0 & 1 \end{pmatrix} \in \mathrm{Mat}_3(\mathbb{R})$$

and the corresponding real quadratic form $Q = -4x_1 x_2 + 2x_1 x_3 - 4x_2^2 + x_3^2$. Let's now make a simultaneous interchange of the first two rows and the first two columns using the following elementary operations

$$A_1 = E_1^{\mathrm{tr}} A E_1 = \begin{pmatrix} -4 & -2 & 0 \\ -2 & 0 & 1 \\ 0 & 1 & 1 \end{pmatrix} \qquad \text{where} \qquad E_1 = \begin{pmatrix} 0 & 1 & 0 \\ 1 & 0 & 0 \\ 0 & 0 & 1 \end{pmatrix}$$

Now let's do another elementary operation

$$A_2 = E_2^{\mathrm{tr}} A_1 E_2 = \begin{pmatrix} -4 & 0 & 0 \\ 0 & 1 & 1 \\ 0 & 1 & 1 \end{pmatrix} \qquad \text{where} \qquad E_2 = \begin{pmatrix} 1 & -\frac{1}{2} & 0 \\ 0 & 1 & 0 \\ 0 & 0 & 1 \end{pmatrix}$$

and yet another elementary operation

$$A_3 = E_3^{\mathrm{tr}} A_2 E_3 = \begin{pmatrix} -4 & 0 & 0 \\ 0 & 1 & 0 \\ 0 & 0 & 0 \end{pmatrix} \qquad \text{where} \qquad E_3 = \begin{pmatrix} 1 & 0 & 0 \\ 0 & 1 & -1 \\ 0 & 0 & 1 \end{pmatrix}$$

We have finally obtained a diagonal matrix A_3, but we would prefer to have, on the main diagonal, first the positive entries, then the negative one and lastly the zero entries. In our case it would be enough to simultaneously interchange both the first two rows and first two columns. Let's do that

$$A_4 = E_4^{\mathrm{tr}} A_3 E_4 = \begin{pmatrix} 1 & 0 & 0 \\ 0 & -4 & 0 \\ 0 & 0 & 0 \end{pmatrix} \qquad \text{where} \qquad E_4 = \begin{pmatrix} 0 & 1 & 0 \\ 1 & 0 & 0 \\ 0 & 0 & 1 \end{pmatrix}$$

and we get

$$(E_1 E_2 E_3 E_4)^{\mathrm{tr}} A (E_1 E_2 E_3 E_4) = \begin{pmatrix} 1 & 0 & 0 \\ 0 & -4 & 0 \\ 0 & 0 & 0 \end{pmatrix}$$

But, the mathematicians are not yet completely satisfied. In fact, they notice that *every positive real number a is a square*, more precisely it is the square of that number which is called its *square root* and is usually written $\sqrt{a}$. So, if $a > 0$ one has $a = (\sqrt{a})^2$ and $-a = -(\sqrt{a})^2$. For example, we have the equalities $\sqrt{4} = 2$, $4 = 2^2$, $-4 = -2^2$ and analogously the equalities $3 = (\sqrt{3})^2$, $-3 = -(\sqrt{3})^2$.

How can we use this observation? Let's look back, for a moment, to the example above and observe that

$$
\begin{pmatrix} 1 & 0 & 0 \\ 0 & -\frac{1}{2} & 0 \\ 0 & 0 & 1 \end{pmatrix}
\begin{pmatrix} 1 & 0 & 0 \\ 0 & -4 & 0 \\ 0 & 0 & 0 \end{pmatrix}
\begin{pmatrix} 1 & 0 & 0 \\ 0 & -\frac{1}{2} & 0 \\ 0 & 0 & 1 \end{pmatrix}
=
\begin{pmatrix} 1 & 0 & 0 \\ 0 & -1 & 0 \\ 0 & 0 & 0 \end{pmatrix}
$$

If we put

$$
E_5 = \begin{pmatrix} 1 & 0 & 0 \\ 0 & -\frac{1}{2} & 0 \\ 0 & 0 & 1 \end{pmatrix}
\qquad
B = \begin{pmatrix} 1 & 0 & 0 \\ 0 & -1 & 0 \\ 0 & 0 & 0 \end{pmatrix}
$$

$$
P = E_1 E_2 E_3 E_4 E_5 = \begin{pmatrix} 1 & 0 & -1 \\ -\frac{1}{2} & -\frac{1}{2} & \frac{1}{2} \\ 0 & 0 & 1 \end{pmatrix}
$$

we have, as a consequence, the equality

$$
P^{\mathrm{tr}} A P = B
$$

Notice that not only is B diagonal, but it has the extra feature that the entries on the diagonal are numbers in the set $\{1, 0, -1\}$ which are, moreover, arranged in order, in the sense that we first find a sequence of 1's, then a sequence of -1's and finally a sequence of 0's. A matrix with these characteristics will be called **a matrix in canonical form**. If we put $P = M_F^E$, we get $F = (v_1, v_2, v_3)$ where $v_1 = (1, -\frac{1}{2}, 0)$, $v_2 = (0, -\frac{1}{2}, 0)$, $v_3 = (-1, \frac{1}{2}, 1)$. Given that $\det(P) = -\frac{1}{2}$, the matrix P is invertible and hence F is a basis for $\mathbb{R}^3$. If we set $(x_1, x_2, x_3)^{\mathrm{tr}} = M_v^E$, $(y_1, y_2, y_3)^{\mathrm{tr}} = M_v^F$ we have

$$
(x_1, x_2, x_3)^{\mathrm{tr}} = M_v^E = M_F^E M_v^F = P (y_1, y_2, y_3)^{\mathrm{tr}}
$$

and hence

$$
\begin{pmatrix} x_1 \\ x_2 \\ x_3 \end{pmatrix} =
\begin{pmatrix} 1 & 0 & -1 \\ -\frac{1}{2} & -\frac{1}{2} & \frac{1}{2} \\ 0 & 0 & 1 \end{pmatrix}
\begin{pmatrix} y_1 \\ y_2 \\ y_3 \end{pmatrix} =
\begin{pmatrix} y_1 - y_3 \\ -\frac{1}{2}y_1 - \frac{1}{2}y_2 + \frac{1}{2}y_3 \\ y_3 \end{pmatrix}
$$

If we make that substitution into the expression $Q = -4x_1 x_2 + 2x_1 x_3 - 4x_2^2 + x_3^2$ we obtain the equality

$$
4(y_1 - y_3)(-\tfrac{1}{2}y_1 - \tfrac{1}{2}y_2 + \tfrac{1}{2}y_3) + 2(y_1 - y_3)y_3 - 4(-\tfrac{1}{2}y_1 - \tfrac{1}{2}y_2 + \tfrac{1}{2}y_3)^2 + y_3^2 =
$$
$$
y_1^2 - y_2^2
$$

As expected from the earlier calculations, one has $M_Q^F = B$. As we did earlier, one says that $Q = y_1^2 - y_2^2$ is the **canonical form of the quadratic form** Q. It shouldn't surprise you that all that we did on this specific example is actually quite general and we have the following fact.

> **Every real symmetric matrix is congruent to a matrix in canonical form.**

Equivalently, one has the following fact.

> **Every real quadratic form may be put into canonical form.**

The sense of all of this is that there is a basis F of $\mathbb{R}^n$ such that if we express the quadratic form Q in coordinates with respect to this basis then Q can be written in the following way

$$y_1^2 + \cdots + y_r^2 - y_{r+1}^2 - \cdots - y_{r+s}^2 \quad \text{with} \quad r + s \leq n$$

5.3 Quadratic Forms, Functions, Positivity

Now that we have seen how to transform the matrix representing a quadratic form into a diagonal matrix (indeed, into a matrix in canonical form) this allows us to study some important properties of quadratic forms. In particular we are interested in studying the behavior of a real quadratic form *thought of as a function from $\mathbb{R}^n$ to $\mathbb{R}$*. This is something that we haven't considered up to now, so this is a good time to pause and reflect a bit on this new way of thinking about quadratic forms. Let's begin by looking at an example. Consider the polynomial $f = x^2 - yz^3 + x - 1$. If, in place of x, y, z, we put some real numbers then, using the indicated operations, we get a real number. For example, if $x = 2$, $y = \frac{1}{2}$, $z = -5$, and $v = (2, \frac{1}{2}, -5)$, we get $f(v) = \frac{135}{2}$. This shows that the polynomial f can act like a function with input from $\mathbb{R}^3$ and output (or values) in $\mathbb{R}$. Mathematicians describe this very succinctly by saying that one can interpret f as a function from $\mathbb{R}^3$ to $\mathbb{R}$ and they write $f : \mathbb{R}^3 \longrightarrow \mathbb{R}$.

It will not have escaped the notice of the attentive reader that the reasoning above can be applied to any polynomial, in particular to a quadratic form which, as we have said, is nothing but a particular kind of polynomial of degree 2. Thus, if $Q = a_{11}x_1^2 + 2a_{12}x_1x_2 + \cdots a_{nn}x_n^2$ is a quadratic form in n variables, then it can be interpreted as a function $Q : \mathbb{R}^n \longrightarrow \mathbb{R}$. Let's explore this aspect of Q. A first observation is that if v is the zero vector then $Q(v) = 0$. But, if v is not zero can we say anything about $Q(v)$? For example, can we tell if $Q(v) \geq 0$ or $Q(v) \leq 0$? Clearly if we take a specific vector v it's enough to simply calculate $Q(v)$. But, if we want to know this more generally? For example, suppose we want to know if $Q(v) > 0$ for every $v \neq 0$, how could we do that? Certainly we cannot try out an infinite number of specific v's, so we need some other information.

Before going on let's take a small digression of an exquisitely technical nature. A mathematician likes to make clear that to speak of the positivity of a quadratic form, the form has to be defined over an **ordered field**. For example we cannot speak of this concept if the field is $\mathbb{C}$, the field of complex numbers or if the field is $\mathbb{Z}_2$, another field already seen and used in Section 2.5. But, we had already decided at the beginning of this section to work only over the field $\mathbb{R}$ and hence we have no problem, because every non-zero real number is either positive or negative. As usual, to put things into better focus let's look at some examples.

Example 5.3.1. Consider the variables x_1, x_2, x_3, and let $v = (x_1, x_2, x_3)$ be the generic vector of $\mathbb{R}^3$. The quadratic form

$$Q(v) = x_1^2 + x_2^2 + 3x_3^2 \tag{1}$$

has the property that $Q(v) > 0$ for every vector $v \neq 0$. In fact, the square of any non-zero real number is positive and thus, when we substitute (in Q) for the three coordinates of the vector, the three summands assume non-negative values. On the other hand, at least one of the three summands is positive, given that the vector is different from the zero vector and hence has at least one coordinate that is not zero.

The quadratic form

$$Q(v) = 3x_1^2 + 8x_3^2 \tag{2}$$

has the property that $Q(v) \geq 0$ for every vector $v \neq 0$. It is different from the preceding example in that it assumes the value zero even for some vectors that are not equal to the zero vector, for example at the vector $v = (0, 1, 0)$.

The quadratic form

$$Q(v) = 3x_1^2 - 8x_3^2 \tag{3}$$

assumes both positive and negative values. For example, we have: $Q(1, 0, 0) = 3$, $Q(0, 0, 1) = -8$.

The reader will probably have noticed that the examples above were easy to analyze because the quadratic forms considered all had associated symmetric matrices which were diagonal. Having an associated symmetric matrix that is diagonal means that the coefficients of the mixed terms, i.e. of the terms $x_i x_j$ with $i \neq j$, were all zero. Or, as we say in a more colloquial way, the mixed terms weren't there. But, consider the quadratic form

$$Q(v) = 2x_1^2 + 2x_1 x_2 + 2x_2^2 + 2x_2 x_3 + 3x_3^2 \tag{4}$$

What can we say about it?

We have to be a bit shrewd here. If, apart from the canonical basis E, we consider another basis F, we know that every vector v can be represented using either E or F and from those two possibilities we get the formulas

$$Q(v) = (M_v^E)^{\mathrm{tr}} M_Q^E M_v^E \qquad Q(v) = (M_v^F)^{\mathrm{tr}} M_Q^F M_v^F$$

Moreover, we also have at our disposal formula (4) of Section 5.1, namely

$$M_Q^F = (M_F^E)^{\text{tr}} M_Q^E M_F^E$$

Now, when we represent the quadratic form Q using M_Q^E or M_Q^F, we get completely different looking representations *of the same quadratic form*. On the other hand, it should be clear that an intrinsic property of the quadratic form *should not depend on its representation*. The problem of positivity concerns a property that is intrinsic to the form, since it depends on how we view the quadratic form as a function. That doesn't depend on how the form is represented and, as a consequence, to study positivity we can use *any matrix M_Q^F*. We've seen above, with the examples, that the absence of mixed terms makes solving the problem of positivity easier and so a winning strategy is to look for a basis F such that M_Q^F is diagonal, something which we know how to do.

Let's set a bit of terminology. A quadratic form which assumes positive values for every non-zero vector will be called **positive definite**. If, instead, it assumes only non-negative values on non-zero vectors we will call the form **positive semidefinite**. Finally, if the form assumes both positive and negative values we will say that the form is not definite. One can naturally transfer this terminology to real symmetric matrices given that such a matrix can always be thought of as M_Q^E and so defines a quadratic form. For example, we say that the matrix $A = \begin{pmatrix} 0 & 0 \\ 0 & 1 \end{pmatrix}$ is positive semidefinite. In fact we can interpret A as M_Q^E, where $Q = x_2^2$ is a quadratic form in two variables which assumes only non-negative values for every vector of $\mathbb{R}^2$, but assumes the value 0 at vectors which are not the zero vector, like $(1, 0)$.

We still have not answered the question above, i.e. what can we say about the form (4)? We can notice that

$$2x_1^2 + 2x_1x_2 + 2x_2^2 + 2x_2x_3 + 3x_3^2 = x_1^2 + (x_1 + x_2)^2 + (x_2 + x_3)^2 + 2x_3^2$$

from which it follows that the form is, at least, positive semidefinite since it is a sum of squares. On the other hand $Q(v) = 0$ implies that $x_1 = x_1 + x_2 = x_1 + x_3 = x_3 = 0$, from which we deduce that $x_1 = x_2 = x_3 = 0$ and hence that v is the zero vector. In this example we succeeded in answering the question by using a small calculational artifice to notice that Q was positive definite. Naturally, in general, we can't hope to always be able to find such artifices. Fortunately, we have available a method which will allow us to always give an answer to this kind of question.

Recall that we have seen, in the preceding section, that if Q is a quadratic form then, with respect to the right basis, Q can be represented with a diagonal matrix. Thus there is a basis F such that, if we call $y_1, y_2, \ldots, y_n$ the coordinates of the generic vector with respect to F, we can write the form Q as

$$Q(v) = b_{11}y_1^2 + b_{22}y_2^2 + \cdots + b_{nn}y_n^2$$

Once we are at this point it is clear that the positivity of the form simply depends on the signs of the coefficients b_{ij}. More precisely,

If the coefficients b_{ij} are all positive then the form is positive definite; if all the coefficients are non-negative and some are zero then the form is positive semidefinite; if the coefficients have different signs then the form is not definite.

Is that the end of the questions? Is it really necessary to represent the form with a diagonal matrix to study the positivity. Let's try a little experiment.

Example 5.3.2. Let $A = \begin{pmatrix} a & b \\ b & c \end{pmatrix}$, and let $Q = ax_1^2 + 2bx_1x_2 + cx_2^2$ be the corresponding quadratic form and let's further suppose that $a \neq 0$. Using an elementary change we obtain

$$B = E^{\text{tr}} AE = \begin{pmatrix} a & 0 \\ 0 & c - \frac{b^2}{a} \end{pmatrix} \quad \text{where} \quad E = \begin{pmatrix} 1 & -\frac{b}{a} \\ 0 & 1 \end{pmatrix}$$

We notice the following facts: the entry in position $(1,1)$ has not been changed; the determinant has not been changed either, it is equal to $ac - b^2$; if we set $\delta = c - \frac{b^2}{a} = \frac{\det(A)}{a}$, then the form, with respect to a new basis, can be written $ay_1^2 + \delta y_2^2 = ay_1^2 + \frac{\det(A)}{a} y_2^2$. From what was said earlier, the quadratic form is positive definite if $a > 0$ and $\delta > 0$ i.e. if $a > 0$ and $\det(A) > 0$.

Thus, for matrices of type 2 with $a_{11} \neq 0$, the positivity can be decided by observing the entry in position $(1,1)$ and by calculating the determinant. What about more generally? In the meantime the reader could amuse himself or herself by verifying that if $a_{11} = 0$, the matrix cannot be positive definite. If that sort of amusement doesn't appeal to you then you can go on reading what follows.

A minor of type (or order) r of a matrix is the determinant of a submatrix of type r. The ith principal minor of a matrix is the determinant of the submatrix formed by the entries in position (r,s) where $1 \leq r \leq i$, $1 \leq s \leq i$.

Using the reasoning of the previous example, we can prove the so-called **Criterion of Sylvester** which affirms the following fact.

A quadratic form represented by a symmetric matrix A is positive definite if and only if all of its principal minors are positive.

For example, the quadratic form associated to the matrix $A = \begin{pmatrix} 1 & 2 \\ 2 & 7 \end{pmatrix}$ is positive definite because $a_{11} = 1 > 0$ and $\det(A) = 3 > 0$, while that associated to the matrix $A = \begin{pmatrix} 0 & 1 \\ 1 & 7 \end{pmatrix}$ is not positive definite because $a_{11} = 0$.

I would like to close this section with this *mathematical gem*. From Sylvester's criterion we can deduce that the positivity of the principal minors of the matrix of a quadratic form does not depend on the basis chosen to represent the form. This is not a trivial fact.

5.4 Cholesky Decomposition

In the preceding section we concentrated on *studying* the positivity of a quadratic form. But, if instead the problem were that of *constructing* positive definite quadratic forms (or positive semidefinite quadratic form), how would we deal with that? One solution is already available and we have seen it in the preceding section; write a quadratic form using a diagonal matrix having all elements on the principal diagonal positive (or all non-negative).

But, there is another way to get positive definite (or positive semidefinite) quadratic forms which is also interesting and which creates symmetric matrices which may not be diagonal. Let's see how. Suppose we have any matrix, A, not necessarily even square. Consider its transpose matrix, A^{tr}. Now form the product $A^{\mathrm{tr}}A$. To begin with, if we call B the matrix $A^{\mathrm{tr}}A$, we have

$$B^{\mathrm{tr}} = (A^{\mathrm{tr}}A)^{\mathrm{tr}} = A^{\mathrm{tr}}(A^{\mathrm{tr}})^{\mathrm{tr}} = A^{\mathrm{tr}}A = B$$

and hence B is symmetric of type c and so we can think of it as the matrix of a quadratic form Q in c variables.

Now comes the interesting discovery. If $v = M_v^E$ is a generic vector of $\mathbb{R}^c$, we have

$$Q(v) = (M_v^E)^{\mathrm{tr}} B M_v^E = (M_v^E)^{\mathrm{tr}} A^{\mathrm{tr}} A M_v^E = (A M_v^E)^{\mathrm{tr}} (A M_v^E)$$

If we put

$$A M_v^E = \begin{pmatrix} y_1 \\ y_2 \\ \vdots \\ y_r \end{pmatrix} \tag{$*$}$$

we get

$$Q(v) = \begin{pmatrix} y_1 & y_2 & \cdots & y_r \end{pmatrix} \begin{pmatrix} y_1 \\ y_2 \\ \vdots \\ y_r \end{pmatrix} = y_1^2 + y_2^2 + \cdots + y_r^2$$

At this point we can already affirm that the form is positive semidefinite. Can we also decide if it is positive definite? In practice we have to see if it is true that $Q(v) = 0$ implies that $v = 0$. But the equality $Q(v) = 0$ is equivalent to the equality $y_1^2 + y_2^2 + \cdots + y_r^2 = 0$, which (in turn) is equivalent to having $y_i = 0$ for every $i = 1, \ldots, r$. Thus, we have to decide if having $y_i = 0$ for every $i = 1, \ldots, r$ implies $v = 0$. From formula $(*)$ we see that having $y_i = 0$ for each $i = 1, \ldots, r$ is precisely the same as having $A M_v^E = 0$ and that *does not, in general, imply that* $M_v^E = 0$. It does imply that, however, **if** A is invertible.

That's not the only situation that will imply that $v = 0$ (as we will see in Section 6.3) but, nevertheless, it is a very important special case and allows us to affirm the following important fact.

If A is an invertible matrix then $A^{\mathrm{tr}}A$ is a positive definite matrix.

Example 5.4.1. If $A = \left(\begin{smallmatrix} 1 & 1 & 0 \\ 2 & 1 & 3 \end{smallmatrix}\right)$, we have

$$A^{\mathrm{tr}}A = \begin{pmatrix} 5 & 3 & 6 \\ 3 & 2 & 3 \\ 6 & 3 & 9 \end{pmatrix}$$

From the discussion above, $A^{\mathrm{tr}}A$ is positive semidefinite. We can see that it is not positive definite in two different ways.

(1) Let's put $A^{\mathrm{tr}}A$ into diagonal form. We obtain the matrix

$$D = \begin{pmatrix} 5 & 0 & 0 \\ 0 & \frac{1}{5} & 0 \\ 0 & 0 & 0 \end{pmatrix}$$

which is not positive definite since there is a zero entry on the principal diagonal. On the other hand, D is the matrix of the quadratic form Q defined by $M_Q^E = A^{\mathrm{tr}}A$. Thus the form Q is not positive definite and hence $A^{\mathrm{tr}}A$ is not positive definite.

(2) Let's look for a non-zero solution to the system of linear equations $A\mathbf{x} = 0$. For example $(3, -3, -1)$ is such a solution. Given $v = (3, -3, -1)$, we therefore have $AM_v^E = 0$ and so $Q(v) = (AM_v^E)^{\mathrm{tr}}(AM_v^E) = 0$.

A curious, and mathematically relevant, thing is that there exists a kind of inverse to what we saw above. In other words, if A is the matrix of a positive definite form then there is an **upper triangular matrix with positive diagonal** U such that $A = U^{\mathrm{tr}}U$. Such a decomposition of the matrix A is called a **Cholesky decomposition** of A. As is now our usual procedure, let's look at an example.

Example 5.4.2. Consider the symmetric matrix

$$A = \begin{pmatrix} 1 & 3 & 1 \\ 3 & 15 & 1 \\ 1 & 1 & 2 \end{pmatrix}$$

Inasmuch as $a_{11} = 1$, the principal minor of type 2 is 6, and the determinant is 2, we deduce from Sylvester's criterion that A is positive definite. Let's try to find the Cholesky decomposition of A. Recall that we want to find an upper triangular matrix U with positive diagonal such that we have the equality $A = U^{\mathrm{tr}}U$.

Set

$$U = \begin{pmatrix} a & b & c \\ 0 & d & e \\ 0 & 0 & f \end{pmatrix} \quad \text{and then} \quad U^{\mathrm{tr}}U = \begin{pmatrix} a^2 & ab & ac \\ ab & b^2+d^2 & bc+de \\ ac & bc+de & c^2+e^2+f^2 \end{pmatrix}$$

Setting that equal to A and taking account of the fact that the diagonal entries of U are to be positive, we obtain the following equalities: $a = 1$, $b = 3$, $c = 1$, $d = \sqrt{6}$, $e = -\frac{2}{\sqrt{6}}$, $f = \frac{1}{\sqrt{3}}$.

In fact, if we set

$$U = \begin{pmatrix} 1 & 3 & 1 \\ 0 & \sqrt{6} & -\frac{2}{\sqrt{6}} \\ 0 & 0 & \frac{1}{\sqrt{3}} \end{pmatrix}$$

we can verify the equality

$$A = \begin{pmatrix} 1 & 3 & 1 \\ 3 & 15 & 1 \\ 1 & 1 & 2 \end{pmatrix} = \begin{pmatrix} 1 & 0 & 0 \\ 3 & \sqrt{6} & 0 \\ 1 & -\frac{2}{\sqrt{6}} & \frac{1}{\sqrt{3}} \end{pmatrix} \begin{pmatrix} 1 & 3 & 1 \\ 0 & \sqrt{6} & -\frac{2}{\sqrt{6}} \\ 0 & 0 & \frac{1}{\sqrt{3}} \end{pmatrix} = U^{\mathrm{tr}} U$$

And what about the case when the matrix is not positive definite? If the reader has paid attention to the discussion above, the following reasoning will not be difficult to follow. If we have the equality $A = U^{\mathrm{tr}} U$, with U upper triangular having positive principal diagonal, then A is positive semidefinite. Moreover, U is invertible and we have recently seen that if U is invertible then $U^{\mathrm{tr}} U$ is, necessarily, positive definite. The conclusion is that if A is not positive definite it cannot have a Cholesky decomposition. Let's look at an example.

Example 5.4.3. Let $A = \begin{pmatrix} 1 & 1 \\ 1 & 0 \end{pmatrix}$. We put $U = \begin{pmatrix} a & b \\ 0 & c \end{pmatrix}$ and set $A = U^{\mathrm{tr}} U$. Doing the calculations we obtain $a^2 = 1$, $ab = 1$, $b^2 + c^2 = 0$, from which we deduce that $a = 1$, $b = 1$ and also $1 + c^2 = 0$, which has no real solution. Some more informed reader would correctly observe that the system is solvable over the complex numbers. But, one shouldn't get overly excited by that fact inasmuch as the field of complex numbers is not an ordered field and so the notion of positivity is not relevant, as we observed at the beginning of Section 5.3.

To finish off this section, let's look at a pair of *mathematical observations*. The first observations is that even if the entries of the positive definite matrix are rational, the Cholesky decomposition may introduce square roots of rational numbers and hence, as in the example just above, the decomposition can be done if one is willing to allow real entries that may not be rational. With that proviso we have no problem with rational matrices.

The second observation is the following. Perhaps you have wondered about how to find, in general, the Cholesky decomposition of a positive definite matrix A. In other words, how does one go about proving that such a decomposition exists in general and how does one find it. As has been said many times, this is the job of a mathematician. If the reader does not have this mathematical curiosity he or she can skip the final part of this section. But, at this stage perhaps every reader has a *little* curiosity?

So, we now switch to *mathematical mode* and prove that every positive definite matrix A has a Cholesky decomposition.

If A is a positive definite matrix then we know that all of its principal minors are positive. Inasmuch as elementary transformations do not change principal minors (a not so obvious fact), after every elementary operation on the rows and columns we find ourselves with a non-zero pivot. Hence we can proceed with elementary transformations (with no exchanges of rows or columns) and arrive at the diagonal form. Then we have the formula $D = P^{\mathrm{tr}} A P$, with D diagonal and all the diagonal entries positive. The matrix P is a product of upper triangular matrices whose diagonal entices are all equal to 1, hence P is an upper triangular matrix whose diagonal entries are all 1. The inverse of P is also, then, upper triangular with diagonal entries all equal to 1. We can write $A = (P^{-1})^{\mathrm{tr}} D P^{-1}$. Now observe that the entries on the diagonal of D are all positive and hence are squares. If we indicate by $\sqrt{D}$ the diagonal matrix whose diagonal entries are the positive square roots of the corresponding entries of D, we obtain $A = (P^{-1})^{\mathrm{tr}} \sqrt{D}^{\mathrm{tr}} \sqrt{D} P^{-1}$. Set $U = \sqrt{D} P^{-1}$ and this gives rise to the conclusion that $A = U^{\mathrm{tr}} U$ and we see that U has the properties we required.

As you can see, *proving* is not an easy job and is justly left to mathematicians, but it is important that the reader understand that it is crucial that someone does this job, otherwise we would just continue to accumulate examples but would never be sure that we were able to affirm something general.

Exercises

Exercise 1. Which of the following expressions are quadratic forms?

(a) $x^2 - 1$
(b) xyz
(c) $x^3 - y^3 + xy - (x - y)^3 - 3xy(x - y) - y^2$

Exercise 2. For which values of a is the quadratic form $x^2 - axy - a^2 y^2$ the square of a linear form?

Exercise 3. Given the following symmetric matrix

$$A = \begin{pmatrix} -4 & 2 \\ 2 & 5 \end{pmatrix}$$

(a) Transform A into a diagonal matrix B using only elementary operations on the rows and columns of A.
(b) Describe the new coordinate system for which the quadratic form Q, associated to the matrix A, now has associated matrix B, and verify your choice.
(c) Deduce that Q is not positive definite by finding two vectors u and v in $\mathbb{R}^2$ such that $Q(u) > 0$ and $Q(v) < 0$.
(d) Do there exist non-zero vectors $u \in \mathbb{R}^2$ such that $Q(u) = 0$?

Exercise 4. *(Hard)*
Let Q be the quadratic form, defined over $\mathbb{Z}_2$, having associated matrix

$$M_Q^E = \begin{pmatrix} 0 & 1 \\ 1 & 0 \end{pmatrix}$$

Show that there is no basis F of $(\mathbb{Z}_2)^2$ for which M_Q^F is diagonal.

Exercise 5. Let Q be the quadratic form defined by

$$M_Q^E = \begin{pmatrix} 0 & -1 & 4 \\ -1 & 0 & -2 \\ 4 & -2 & 7 \end{pmatrix}$$

and let $f_1 = (1, 1, 1)$, $f_2 = (1, 0, 2)$, $f_3 = (0, 2, -1)$ and $F = (u_1, u_2, u_3)$.

(a) Verify that F is a basis of $\mathbb{R}^3$.
(b) Calculate M_Q^F.

Exercise 6. Consider the following symmetric matrix

$$A = \begin{pmatrix} 8 & -4 & 4 \\ -4 & 6 & 18 \\ 4 & 18 & 102 \end{pmatrix}$$

(a) Decide if A is positive definite, positive semidefinite or not definite.
(b) Using elementary row and column operations, transform A into a diagonal matrix B.
(c) Determine all the vectors $u \in \mathbb{R}^3$ such that $Q(u) = 0$.

Exercise 7. Given the following matrix

$$A = \begin{pmatrix} 1 & 2 & 3 \\ 2 & 4 & 6 \\ 4 & 8 & 12 \end{pmatrix},$$

(a) Let Q be the quadratic form associate to the matrix $B = A^{\mathrm{tr}} A$. Put Q in canonical form and find the change of basis matrix.
(b) Is it true that Q is the square of a linear form?

Exercise 8. Consider the set S of real symmetric matrices of type 3 (or, as the mathematicians would say; consider the subset $S \subset \mathrm{Mat}_3(\mathbb{R})$ of symmetric matrices).

(a) How many matrices in S are in canonical form? What are they?
(b) How many matrices in S are in canonical form and are also positive semidefinite? What are they?

Exercise 9. Let $A \in \mathrm{Mat}_n(\mathbb{R})$ be a real symmetric matrix.
(a) Is it true that if A is positive definite then $a_{ii} > 0$ for $i = 1, \ldots, n$?
(b) Is the converse also true?

Exercise 10. Given the matrix

$$A = \begin{pmatrix} 1 & 2 \\ 2 & 4 \\ 4 & 0 \end{pmatrix}$$

(a) Find the Cholesky decomposition of $A^{\mathrm{tr}} A$.
(b) Is it possible to find the Cholesky decomposition of AA^{tr}?

@ **Exercise 11.** Put the following symmetric matrix into canonical form

$$A = \begin{pmatrix} 2 & 3 & 4 & 2 & 2 \\ 3 & 1 & 7 & 1 & 6 \\ 4 & 7 & 10 & 4 & 6 \\ 2 & 1 & 4 & 1 & 3 \\ 2 & 6 & 6 & 3 & 3 \end{pmatrix}$$

6

Orthogonality and Orthonormality

dove sta esattamente l'ortocentro?
quale è la dimensione di un orto normale?

(From "LE DOMANDE DELL'ORTICOLTORE"
by an anonymous author)

We are used to thinking of orthogonal coordinate systems as the most interesting and most useful. This habit comes from studying the graphs of functions in the plane or in space. Is the same thing also true in $\mathbb{R}^n$? In this chapter we will try to justify the reason for this perception and also to respond to the question above.

Our starting point is exactly where we stopped in our study of positive definite quadratic forms in the previous chapter. Among those there is one that is rather special, namely the one for which $M_Q^E = I$. Why is it so special? Recall that, in Example 5.1.3, we saw that we have the relation $Q(v) = (M_v^E)^{\mathrm{tr}} I M_v^E = v \cdot v = |v|^2$ for this form. Thus, this quadratic form is connected to the concept of the length of a vector and from that to the concept of distance. But, there is much more. In fact one can extend this connection between quadratic forms and scalar products quite a bit. We won't go into the details of this rather sophisticated theory; it is enough for the reader to know that quadratic forms are intrinsically tied, through the concept of the *polar form*, to the so-called *bilinear forms* (among which we find the scalar product).

In this chapter we will speak at length about orthogonality and of orthogonal projections. Since we will work in the spaces $\mathbb{R}^n$ we will need some new algebraic instruments. We will have to stop along the way to consider the concepts of *linear dependence, rank of a matrix, vector subspaces and their dimensions*. We will discuss orthogonal and orthonormal matrices and we will see how to construct orthonormal matrices starting from matrices of maximal rank using the so-called *Gram-Schmidt orthonormalization procedure* and the *QR decomposition*. There's lots of work to do!

Robbiano L.: Linear Algebra for everyone
© Springer-Verlag Italia 2011

6.1 Orthonormal Tuples and Orthonormal Matrices

We have already seen in Section 4.6, that the scalar product is the algebraic instrument which allows us to speak of orthogonality even in $\mathbb{R}^n$. Let's suppose that we have an s-tuple of vectors from $\mathbb{R}^n$, $S = (w_1, w_2, \ldots, w_s)$. If we were looking for a clever way to store all the scalar products of the vectors of S, how could we do that? This is not hard; we can consider the matrix M_S^E and recall that the columns of M_S^E contain the coordinates of the vectors of S and that the rows of $(M_S^E)^{\mathrm{tr}}$ coincide with the columns of M_S^E. Thus, if $w_i = (a_1, a_2, \ldots, a_n)$, $w_j = (b_1, b_2, \ldots, b_n)$, then the entry in position (i, j) of the matrix $(M_S^E)^{\mathrm{tr}} M_S^E$ is precisely $a_1 b_1 + a_2 b_2 + \cdots + a_n b_n$. This number is nothing more than the scalar product $w_i \cdot w_j$. We can thus say that *the entry in the (i, j) position of the matrix $(M_S^E)^{\mathrm{tr}} M_S^E$ is $w_i \cdot w_j$* and we can also say that the matrix $(M_S^E)^{\mathrm{tr}} M_S^E$ *is the matrix of the scalar products of the vectors in S*. One of the first consequences of this observation is that the vectors of the s-tuple S are pairwise orthogonal if and only if $(M_S^E)^{\mathrm{tr}} M_S^E$ is a diagonal matrix, and they are pairwise orthogonal and of unit length if and only if $(M_S^E)^{\mathrm{tr}} M_S^E = I_s$. In the first case we say that S is an **orthogonal** s-tuple of vectors and that the matrix M_S^E is an **orthogonal matrix**. In the second case we say that S is an **orthonormal** s-tuple of vectors and that the matrix M_S^E is an **orthonormal matrix**. Notice that in the second case the vectors, having unit length, are automatically not zero. In particular if S is a basis we can speak of an **orthogonal basis** in the first case and of an **orthonormal basis** in the second case. Let's look at an example.

Example 6.1.1. Let $w_1 = (1, 2, 1)$, $w_2 = (1, -1, 1)$ and let $S = (w_1, w_2)$. Then S is an orthogonal but not orthonormal pair of vectors. Equivalently, the matrix

$$M_S^E = \begin{pmatrix} 1 & 1 \\ 2 & -1 \\ 1 & 1 \end{pmatrix}$$

is an orthogonal matrix but not an orthonormal matrix. We see that

$$(M_S^E)^{\mathrm{tr}} (M_S^E) = \begin{pmatrix} 1 & 2 & 1 \\ 1 & -1 & 1 \end{pmatrix} \begin{pmatrix} 1 & 1 \\ 2 & -1 \\ 1 & 1 \end{pmatrix} = \begin{pmatrix} 6 & 0 \\ 0 & 3 \end{pmatrix}$$

is a diagonal matrix but is not the identity matrix.

We have arrived at a central point. What characterizes a matrix of type M_S^E when S is an orthonormal basis? As we have just seen, if $A = M_S^E$, where S is an orthonormal basis, then $A^{\mathrm{tr}} A = I$. But, we already know that A is invertible and thus we get the relation

$$A^{\mathrm{tr}} = A^{-1}$$

In other words, *its inverse coincides with its transpose!* The very attentive reader will have noticed that once we have the relation $A^{\mathrm{tr}} A = I$, then in order to conclude that $A^{\mathrm{tr}} = A^{-1}$ it is enough to know that A is a square matrix (see Section 2.5). The interesting consequence of this observation is the following

Every orthonormal n-tuple of vectors of $\mathbb{R}^n$ is a basis for $\mathbb{R}^n$.

Example 6.1.2. Let's consider the following matrix

$$A = \begin{pmatrix} 1 & 0 & 0 \\ 0 & \frac{1}{\sqrt{2}} & -\frac{1}{\sqrt{2}} \\ 0 & \frac{1}{\sqrt{2}} & \frac{1}{\sqrt{2}} \end{pmatrix}$$

This is an orthonormal matrix, in fact we see easily that $A^{\mathrm{tr}} A = I$, or, equivalently that the three vectors whose coordinates form the columns of A, are pairwise orthogonal and of unit length. As a consequence, the triple of vectors whose coordinates constitute the columns of A is an orthonormal basis for $\mathbb{R}^3$.

We conclude this section with an observation. In the case of square orthonormal matrices the *very expensive* operation of calculating the inverse is reduced to the *very cheap* operation of calculating the transpose. This reveals a big reason why orthonormal matrices are considered so important.

6.2 Rotations

> *l'altra luna faccia la rivoluzione*
> *e mostri l'una e l'altra faccia*
>
> (From "ROTAZIONI E RIVOLUZIONI"
> by an anonymous author)

Let's concentrate for the moment on the orthogonal matrices of type 2 and try to classify them. First of all, what does it mean to classify these objects? In the mathematical context it means much the same as it would mean in other contexts, namely to subdivide the objects on the basis of some predetermined characteristics. Naturally, the phrase is still rather vague and we will start this classification with the matrices of type 2 and see if we can get some idea of what we might mean. Notice that such a matrix may be written in the following way

$$O = \begin{pmatrix} a & b \\ c & d \end{pmatrix}$$

where the condition of orthogonality gives

$$ab + cd = 0 \tag{1}$$

while that of normality gives

$$a^2 + c^2 = 1 \qquad b^2 + d^2 = 1 \tag{2}$$

Observe that two real numbers a and c such that $a^2 + c^2 = 1$ are necessarily the cosine and sine of the same angle ϑ. So, we can assume that $a = \cos(\vartheta)$ and $c = \sin(\vartheta)$. The same thing is true for b and d. Thus we can assume that $b = \cos(\varphi)$ and $d = \sin(\varphi)$. On the other hand, the orthogonality of the two vectors, expressed by equation (1), implies the angles ϑ and φ differ by $\frac{\pi}{2}$. Thus $\varphi = \vartheta + \frac{\pi}{2}$ or $\varphi = \vartheta - \frac{\pi}{2}$ and so $\cos(\varphi) = -\sin(\vartheta)$ or $\cos(\varphi) = \sin(\vartheta)$ and $\sin(\varphi) = \cos(\vartheta)$ or $\sin(\varphi) = -\cos(\vartheta)$. Putting this all together one has

$$O = \begin{pmatrix} \cos(\vartheta) & -\sin(\vartheta) \\ \sin(\vartheta) & \cos(\vartheta) \end{pmatrix} \qquad \text{or} \qquad O = \begin{pmatrix} \cos(\vartheta) & \sin(\vartheta) \\ \sin(\vartheta) & -\cos(\vartheta) \end{pmatrix} \tag{3}$$

One notices that the first matrix has determinant 1 while the second matrix has determinant -1. We thus have in front of us a *classification*. In fact we have a description of the family of orthonormal matrices of type 2 divided into two subfamilies. The *tag* which identifies each member of the subfamily is the value of ϑ.

Can we give a geometric significance to each of the two families? The answer to this question is rather simple but requires some preliminary considerations. Do you remember the significance of the determinant of a matrix of type 2? That problem was studied in detail in Section 4.4 where we concluded by saying that *the determinant of a matrix is, in absolute value, the area of the parallelogram defined by the two vectors whose coordinates, with respect to an orthogonal cartesian system of vectors of unit length, are the columns of the matrix*. In our case, since we are dealing with two vectors of unit length (see formula (2)) which are orthogonal to each other (see formula (1)), the parallelogram in question is a square having side of length 1 and hence has area 1.

Now let's reflect on that small additional piece of information in the phrase, *in absolute value*. What does that mean here? Recall that if we exchange two columns of a square matrix its determinant changes sign (see rule (a) in Section 4.6). So, we cannot always assume that the determinant is an area. However, the determinant *contains more information*. It's absolute value is the area, and the sign depends on the direction in which we think of the angle formed by the two vectors corresponding to the columns. If we move from the first column to the second in a counterclockwise sense, the sign is positive, if the motion is in a clockwise sense, the sign is negative. Why? Describing the matrices using formulas (3), as we did before, let's see if we can clear up this rule of signs. Since the vector corresponding to the first column is $(\cos(\vartheta), \sin(\vartheta))$, the vector $(-\sin(\vartheta), \cos(\vartheta))$ coincides with $(\cos(\vartheta + \frac{\pi}{2}),$ $\sin(\vartheta + \frac{\pi}{2}))$, while the vector $(\sin(\vartheta), -\cos(\vartheta))$ coincides with $(\cos(\vartheta - \frac{\pi}{2}),$ $\sin(\vartheta - \frac{\pi}{2}))$. Inasmuch as we conventionally sum angles in the counterclockwise sense, the conclusion is clear.

After all of these considerations it would be good if the reader has understood why this section is called *rotations*.

6.3 Subspaces, Linear Independence, Rank, Dimension

If it's true that orthonormal bases are so important, as we have already noted in the preceding sections, then it is certainly worth the effort to look for them and possibly to construct them. Before coming to grips with this new challenge it will be good to equip ourselves with some more mathematical tools.

Let's recall that $F = (f_1, \ldots, f_n)$ is a basis for $\mathbb{R}^n$ if and only if M_F^E is invertible. Remember also that being a basis essentially points out two important features of F, namely that every vector in $\mathbb{R}^n$ can be written as a linear combination of this n-tuple of vectors and that this way of writing the vector as a linear combination of the n-tuple of vectors is unique.

Suppose now that we have an s-tuple of vectors $G = (g_1, \ldots, g_s)$ with $s < n$. We already know that G cannot be a basis for $\mathbb{R}^n$ because G doesn't have enough vectors in it. But, couldn't it be a basis for a smaller space or, more modestly, generate a smaller space? Now here someone had a great idea, namely considering the *space $V(G)$ consisting of all the vectors which are linear combinations of $g_1, \ldots, g_s$*. Mathematicians call this the **vector subspace of $\mathbb{R}^n$ generated by G**, while $\mathbb{R}^n$ is called the **vector space of the n-tuples of real numbers**. If it is also true that every vector in $V(G)$ can be written uniquely as a linear combination of the vectors of G then we say that G is an s-tuple of *linearly independent vectors* or that G **is a basis for $V(G)$**. Moreover, if G is an orthogonal (or orthonormal) s-tuple we say that G is an **orthogonal (orthonormal) basis** of $V(G)$. Let's try to get some familiarity with these ideas and the new terminology by looking at a geometric example.

Example 6.3.1. Let $g_1 = (1, 1, 1)$, $g_2 = (-1, 2, 0)$, $g_3 = (1, 4, 2)$ be three vectors in $\mathbb{R}^3$ and let $G = (g_1, g_2, g_3)$. Consider the matrix M_G^E and the system of linear homogeneous equations associated to them

$$\begin{cases} x_1 - x_2 + x_3 = 0 \\ x_1 + 2x_2 + 4x_3 = 0 \\ x_1 \quad\quad + 2x_3 = 0 \end{cases}$$

When we solve this system we will find an infinite number of solutions, among them $(2, 1, -1)$. In fact, the given solution corresponds to the relation $2g_1 + g_2 - g_3 = 0$. Since the zero vector can also be written in this way $0g_1 + 0g_2 + 0g_3 = 0$ one sees that the zero vector can be written in more than one way as a linear combination of the vectors of G. This indicates that the columns of M_G^E are *linearly dependent* or, equivalently, that the three vectors of G are linearly dependent i.e. not linearly independent.

Now let's consider the space $V(G)$ which consists of all the linear combinations of the vectors in G. The relation $2g_1 + g_2 - g_3 = 0$ can also be read as $g_3 = 2g_1 + g_2$. Thus, the vector g_3 is a linear combination of g_1 and g_2 and thus if we call G' the pair (g_1, g_2) we have that $V(G) = V(G')$. If we now solve the homogeneous system of linear equations associated to $M^E_{G'}$, we see that it has only the trivial solution and hence the vectors of G' are linearly independent and hence G' is a basis for the vector subspace $V(G) = V(G')$. Let's now suppose that we have a system of Cartesian coordinates in space and let's interpret g_1, g_2 and g_3 as vectors (or points) in the way described in Section 4.2. Geometrically speaking, we can say that the two non-parallel vectors g_1 and g_2, generate a plane π passing through the origin of the coordinate system and that the third vector g_3 is in π. Moreover, the two vectors g_1 and g_2 with the origin O constitute a system of coordinates $\Sigma(O; g_1, g_2)$ on π.

By studying examples like the one above, mathematicians became aware that what happens in that example is not an isolated phenomenon. More precisely, they showed that in order to be sure that G is an r-tuple of linearly independent vectors it is enough to be able to write the zero vector in a unique way, i.e. only as $0\,g_1 + 0\,g_2 + \cdots + 0\,g_s$. Therefore if we consider the matrix M^E_G and the homogeneous system of linear equations associated to it, i.e. $M^E_G\,\mathbf{x} = 0$, to say that G is formed of linearly independent vectors is the same as saying that this system of equations has only the trivial solution. This fact can also be expressed by saying that *the columns of the matrix M^E_G are linearly independent.*

Given any s-tuple of vectors whatsoever, we can ask what is the maximum number of vectors in the s-tuple which are linearly independent. In an analogous way, we can ask what is the maximum number of linearly independent columns in an arbitrary matrix.

Mathematicians may, at times, be quite boring but they are undoubtedly often acute observers and know how to furnish an interesting and complete answer to this question. They have proved the following facts.

(1) **The maximum number of linearly independent columns of a matrix is the same as the maximum number of its linearly independent rows.**
(2) **That number coincides with the maximum type of a square submatrix with non-zero determinant.**
(3) **That number doesn't change if we multiply the matrix by any invertible matrix.**

Recalling what was said in Section 5.3, namely that a minor of type (or order) r is the determinant of a submatrix of type r, rule (2) can be rewritten as follows.

(2') **That number coincides with the maximum type (or order) of a non-zero minor.**

From these remarks, we understand the importance of that number. This merits it being given a name: we call that number the **rank** of A and indicate it with rk(A) (in some countries the word **characteristic** is used instead of rank). Let's look at a simple example.

Example 6.3.2. Let's consider the matrix

$$A = \begin{pmatrix} 1 & 2 & 3 \\ -1 & 1 & 0 \\ -1 & 5 & 4 \end{pmatrix}$$

Since the third column is the sum of the first two, the three column vectors are not linearly independent. Another way to see this is to observe that $\det(A) = 0$. Since the submatrix formed using the first two rows and the first two columns has determinant non-zero, we can conclude that rk(A) = 2 and hence that the maximum number of columns (and rows) of A that are linearly independent is precisely 2.

It's clear from the properties above that rk(A) can never exceed either the number of rows of A or the number of columns of A. Put more precisely we have the following fact:

If A is an $m \times n$ matrix then its rank cannot exceed the minimum of m and n.

We call a matrix which has rank equal to the minimum of the number of rows and the number of columns a matrix of **maximal rank**. Note that the matrix of the example just above is not of maximal rank while the matrix $\begin{pmatrix} 1 & 0 & 2 \\ 0 & 0 & 1 \end{pmatrix}$ is.

In Section 4.8 we saw that every basis of $\mathbb{R}^n$ is formed using n vectors. It is very appropriate then, to call n the **dimension of $\mathbb{R}^n$** since in the cases $n = 1$, $n = 2$ and $n = 3$ it corresponds to our intuitive idea of dimension. How can we extend this concept to vector subspaces? In Example 6.3.1 we saw that the space $V(G)$ is a plane and hence it would be logical for it to have dimension 2. It is no accident that the number of vectors in a basis for it (namely G') is 2. I say "no accident" because we can show that not only does $\mathbb{R}^n$ have all its bases with the same number of elements but any vector subspace V of $\mathbb{R}^n$ has all its bases with the same number of vectors in them. It is quite natural to call that number the **dimension of V** and to denote it dim(V).

Since the rank of any matrix $A \in \mathrm{Mat}_{r,c}(\mathbb{R})$ coincides with the maximum number of linearly independent columns of A, we can deduce the following facts.

(1) **The dimension of the subspace V of $\mathbb{R}^n$ generated by the columns of A coincides with rk(A).**
(2) **A basis of V can be obtained by extracting the maximum number of linearly independent columns from A.**

Let's do another example so that we have these concepts clear.

Example 6.3.3. Consider the following vectors in $\mathbb{R}^5$.

$g_1 = (1, 0, 1, 0, 1)$, $g_2 = (-1, -2, -3, 1, 1)$, $g_3 = (5, 8, 13, -4, -3)$,
$g_4 = (8, 14, 6, 1, -8)$, $g_5 = (-17, -42, -11, -3, 31)$.

Let $G = (g_1, g_2, g_3, g_4, g_5)$ and consider $V(G)$, a vector subspace of $\mathbb{R}^5$. Observe that M_G^E is a square matrix of type 5. In fact one has

$$M_G^E = \begin{pmatrix} 1 & -1 & 5 & 8 & -17 \\ 0 & -2 & 8 & 14 & -42 \\ 1 & -3 & 13 & 6 & -11 \\ 0 & 1 & -4 & 1 & -3 \\ 1 & 1 & -3 & -8 & 31 \end{pmatrix}$$

If we do some elementary row operations, following the strategy of Gaussian reduction, we obtain the matrix

$$A = \begin{pmatrix} 1 & -1 & 5 & 8 & -17 \\ 0 & -2 & 8 & 14 & -42 \\ 0 & 0 & 0 & -16 & 48 \\ 0 & 0 & 0 & 0 & 0 \\ 0 & 0 & 0 & 0 & 0 \end{pmatrix}$$

As A has two zero rows, its rank cannot exceed three. On the other hand we notice that the submatrix formed by the first three rows and by the first, second and fourth column is upper triangular with determinant different from zero. Using rules (1), (2) and (3) about the rank of a matrix, we can thus conclude that $\mathrm{rk}(A) = 3$, that $\mathrm{rk}(A) = \mathrm{rk}(M_G^E)$ and hence that $\dim(V(G)) = 3$. Finally, observe that (g_1, g_2, g_4) is a basis of $V(G)$, while (for example) (g_1, g_2, g_3) is not a basis for $V(G)$, since these last three vectors are linearly dependent.

We conclude this section with a discussion about something which mathematicians (rightly) think is very important. More precisely we will see a class of vector subspaces to which we will return in Section 8.2 when we consider eigenspaces in detail. The key point of this discussion is the fact that *the space of solutions of a system of homogeneous linear equations* is a vector subspace. Let's see why that is true.

Given a number field K (for example $\mathbb{R}$) consider a homogeneous system of linear equations in n unknowns and with coefficients in K. Let V be the set of solutions to that system in K^n. Then the following statements are true.

(1) **The set V is a vector subspace of K^n.**
(2) **After we do Gaussian reduction and then attribute to the free variables the values $(1, 0, \ldots, 0)$, $(0, 1, \ldots, 0)$, $\ldots$, $(0, 0, \ldots, 1)$, the solutions we so obtain form a basis for V.**

Let's clarify these concepts with an example.

Example 6.3.4. Consider the following system of homogeneous linear equations S, with real coefficients,

$$\begin{cases} x_1 - x_2 + 2x_3 - x_4 = 0 \\ x_1 - x_2 + 3x_3 - 4x_4 = 0 \end{cases}$$

With two elementary operations we obtain the following equivalent system

$$\begin{cases} x_1 - x_2 + 5x_4 = 0 \\ x_3 - 3x_4 = 0 \end{cases}$$

Considering x_2 and x_4 as the free variables, the general solution, in $\mathbb{R}^4$, of S is thus $(a - 5b,\ a,\ 3b,\ b)$ where a and b take on arbitrary values in $\mathbb{R}$. Putting $a = 1$, $b = 0$, we get the vector $u_1 = (1, 1, 0, 0)$. Putting $a = 0$, $b = 1$, we get the vector $u_2 = (-5, 0, 3, 1)$.

Using properties (1) and (2) we can conclude that the set V, of real solutions to the homogeneous system S, is a vector subspace of $\mathbb{R}^4$ and that a basis for it is (u_1, u_2).

6.4 Orthonormal Bases and the Gram-Schmidt Procedure

When we used systems of coordinates in the plane, we always preferred to use those defined by a pair of non-zero orthogonal vectors of the same length, i.e. we used coordinates that we called orthogonal and monometric. But, if we already have a system of coordinates, can we construct another with these desirable characteristics?

Let's look at the following figure.

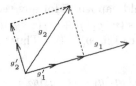

If we had begun with (g_1, g_2) and we had also a unit of measure we could consider the unit vector in the direction of g_1, call it g_1', and decompose g_2 as the sum of two vectors, one parallel to g_1 and one orthogonal to g_1. If we let g_2' be the unit vector in the direction orthogonal to g_1 we see that (g_1', g_2') are orthonormal vectors and hence (together with an origin) define a system of orthogonal coordinates which is monometric. Can we hope to generalize this intuitive discussion to $\mathbb{R}^n$? One can always hope; but, fortunately, in this case we can transform that hope into reality. Let's see how.

We begin with an interesting consideration. Let's suppose that we have at hand an s-tuple $G = (g_1, \ldots, g_s)$ of vectors in $\mathbb{R}^n$ which are linearly independent and are also orthonormal i.e. an orthonormal basis for $V(G)$. Let v be a vector in $V(G)$. We know that we can write $v = GM_v^G = a_1 g_1 + \cdots a_s g_s$. However, if we consider the scalar products $v \cdot g_i$ and use the orthonormality of G we obtain the following relations $v \cdot g_i = (a_1 g_1 + \cdots a_s g_s) \cdot g_i = a_i$. It follows that

$$v = (v \cdot g_1) g_1 + \cdots (v \cdot g_s) g_s \quad \text{i.e.} \quad M_v^G = (v \cdot g_1 \ \cdots \ v \cdot g_s)^{\text{tr}} \quad (*)$$

Let's look at an example.

Example 6.4.1. Consider the pair of vectors $G = (g_1, g_2)$ in the space $\mathbb{R}^4$, where $g_1 = (1, 1, 0, -1)$, $g_2 = (-1, 0, 0, -1)$. Since $g_1 \cdot g_2 = 0$ these are orthogonal vectors. If we consider their normalizations we obtain a pair of orthonormal vectors and hence an orthonormal basis for $V(G)$. Letting $g_1' = \text{vers}(g_1) = \frac{1}{\sqrt{3}}(1, 1, 0, -1) = (\frac{1}{\sqrt{3}}, \frac{1}{\sqrt{3}}, 0, -\frac{1}{\sqrt{3}})$ and $g_2' = \text{vers}(g_2) = \frac{1}{\sqrt{2}}(-1, 0, 0, -1) = (-\frac{1}{\sqrt{2}}, 0, 0, -\frac{1}{\sqrt{2}})$, we see that $G' = (g_1', g_2')$ is an orthonormal basis for $V(G) = V(G')$.

Now, consider the vector $v = g_1 + 2g_2 = (-1, 1, 0, -3)$ in the space $V(G)$. Since $v \cdot g_1' = \sqrt{3}$ and $v \cdot g_2' = 2\sqrt{2}$, we have $(v \cdot g_1') g_1' + (v \cdot g_2') g_2' = \sqrt{3}(\frac{1}{\sqrt{3}}, \frac{1}{\sqrt{3}}, 0, -\frac{1}{\sqrt{3}}) + 2\sqrt{2}(-\frac{1}{\sqrt{2}}, 0, 0, -\frac{1}{\sqrt{2}}) = (1, 1, 0, -1) + (-2, 0, 0, -2) = (-1, 1, 0, -3) = v$, as we expected from formula $(*)$.

The attentive reader will have noticed that formula $(*)$ didn't need the a priori knowledge that the vectors were linearly independent because orthonormality implies independence.

Every s-tuple of orthonormal vectors in $\mathbb{R}^n$ is necessarily a set of linearly independent vectors, and consequently $s \leq n$.

Naturally formula $(*)$ is true for the vectors of $V(G)$, but now comes the question which stimulates another good idea. Let's suppose that the s-tuple G is orthonormal. What would happen if, for any vector v of $\mathbb{R}^n$ we associated another obtained as follows $(v \cdot g_1) g_1 + \cdots (v \cdot g_s) g_s$? First of all, let's give a name to this vector: we'll call it $p_{V(G)}(v)$. Thus, we say that for any vector $v \in \mathbb{R}^n$ we have

$$p_{V(G)}(v) = (v \cdot g_1) g_1 + (v \cdot g_2) g_2 + \cdots + (v \cdot g_s) g_s$$

If G is an s-tuple of orthogonal vectors which are linearly independent (but not necessarily orthonormal) it is not difficult to see that, for every vector $v \in \mathbb{R}^n$, the vector $p_{V(G)}(v)$ is obtained in the following way:

$$p_{V(G)}(v) = \frac{1}{|g_1|^2}(v \cdot g_1) g_1 + \frac{1}{|g_2|^2}(v \cdot g_2) g_2 + \cdots + \frac{1}{|g_s|^2}(v \cdot g_s) g_s$$

Given the importance of the vector $p_{V(G)}(v)$ we will call it the **orthogonal projection of v onto $V(G)$**. Why is it important? Mathematicians have shown that taking $w = v - p_{V(G)}(v)$ we have $w \cdot g_i = 0$ for each $i = 1, \ldots, s$ and hence w is orthogonal to all the vectors of $V(G)$. They have also shown that the vector $p_{V(G)}(v)$ is, among all the vectors of $V(G)$, the **unique vector whose distance from v is the minimum possible.** An important consequence of this fact is that the vector $p_{V(G)}(v)$ does not depend on G but will give the same result for any orthonormal basis of $V(G)$. In conclusion, by means of the construction of $p_{V(G)}(v)$ we have been able to generalize the geometric considerations that we made at the beginning of this section.

We would now like to satisfy those, among the readers, who are curious to know how one might prove the assertions of this last paragraph.

For conciseness, let's set $u = p_{V(G)}(v)$. First we prove that

$$v - u \text{ is orthogonal to all the vectors of } V(G) \tag{1}$$

Using the linearity of the scalar product it is enough to prove that

$$(v - u) \cdot g_i = 0 \text{ for every } i = 1, \ldots, s \tag{2}$$

In fact, $(v-u) \cdot g_i = v \cdot g_i - \sum_{j=1}^{s}(v \cdot g_j)(g_j \cdot g_i)$. In the sum the addenda are all equal to zero, except for the i-th which is equal to $v \cdot g_i$ and so (2) is proved and hence (1) is proved. As a first consequence we have that

$$\text{if } u' \in V(G) \text{ then } v \cdot u' = u \cdot u' \tag{3}$$

In fact, we have $v = (v - u) + u$ and the conclusion follows from the linearity of the scalar product and from (1). In particular one has

$$v \cdot u = |u|^2 \tag{4}$$

Now we can prove the main point, i.e. if $u' \in V(G)$ is any vector then

$$|v - u|^2 \leq |v - u'|^2 \text{ with strict inequality if } u' \neq u.$$

Let's see how to do this. In order to prove that $|v - u|^2 \leq |v - u'|^2$ it is enough to prove that

$$|v|^2 - 2\, v \cdot u + |u|^2 \leq |v|^2 - 2\, v \cdot u' + |u'|^2 \tag{5}$$

Using (3) and (4) we are brought to trying to prove that

$$-2|u|^2 + |u|^2 \leq -2\, u \cdot u' + |u'|^2 \tag{6}$$

or, in other words, that

$$0 \leq |u|^2 - 2\, u \cdot u' + |u'|^2 = |u - u'|^2 \tag{7}$$

Now, equation (7) is true and, even more, it's true that $0 < |u - u'|^2$ if $u' \neq u$. The proof is thus finished.

Let's look at a specific example.

Example 6.4.2. Consider $g_1 = (1, -1, 0)$, $g_2 = (1, 1, 1)$. The two vectors are orthogonal but not of unit length. Thus $G = (g_1, g_2)$ is an orthogonal basis for $V(G)$. Now let $v = (2, 1, -7)$ and let's calculate the orthogonal projection of v on V. We get

$$p_{V(G)}(v) = \tfrac{1}{|g_1|^2}(v \cdot g_1)g_1 + \tfrac{1}{|g_2|^2}(v \cdot g_2)g_2 = \tfrac{1}{2}(1, -1, 0) + \tfrac{1}{3}(-4)(1, 1, 1)$$
$$= (\tfrac{1}{2}, -\tfrac{1}{2}, 0) - \tfrac{4}{3}(1, 1, 1) = (-\tfrac{5}{6}, -\tfrac{11}{6}, -\tfrac{4}{3})$$

At this point we are now ready to introduce a procedure that, starting from an s-tuple G of vectors which are linearly independent, furnishes us with an orthonormal basis of $V(G)$. This procedure is called the **Gram-Schmidt orthonormalization procedure**.

We start with an s-tuple $G = (g_1, \ldots, g_s)$ of linearly independent vectors. The vector g_1 is not the zero vector and hence we can consider its normalization

$$g'_1 = \text{vers}(g_1)$$

and the vector

$$g'_2 = \text{vers}\left(g_2 - p_{V(g'_1)}(g_2)\right) = \text{vers}\left(g_2 - (g_2 \cdot g'_1)g'_1\right)$$

It is easy to check that $g'_1 \cdot g'_2 = 0$. Moreover g'_1, g'_2 are two vectors of unit length and, as we have constructed them, it is also easy to see that $V(g_1, g_2) = V(g'_1, g'_2)$. We can continue in this way up to the last vector, which has the form

$$g'_s = \text{vers}\left(g_s - p_{V(g'_1, \ldots, g'_{s-1})}(g_s)\right) = \text{vers}\left(g_s - (g_s \cdot g'_1)g'_1 - \cdots - (g_s \cdot g'_{s-1})g'_{s-1}\right)$$

In this way we obtain a new basis $G' = (g'_1, \ldots, g'_s)$ of $V(G)$ which, by construction, is orthonormal. Let's see an explicit example.

Example 6.4.3. Let $G = (g_1, g_2)$ where $g_1 = (1, 1, 0)$, $g_2 = (1, 2, 1)$. The two vectors are linearly independent and thus are a basis for $V(G)$. But, they certainly are not an orthonormal basis. Let's apply the Gram-Schmidt procedure. We construct the vector

$$g'_1 = \text{vers}(g_1) = \frac{1}{\sqrt{2}}(1, 1, 0) = (\frac{\sqrt{2}}{2}, \frac{\sqrt{2}}{2}, 0)$$

Now we construct the vector

$$g_2 - (g_2 \cdot g'_1)g'_1 = (1, 2, 1) - \frac{3}{\sqrt{2}}\frac{1}{\sqrt{2}}(1, 1, 0) = (1, 2, 1) - \frac{3}{2}(1, 1, 0) = (-\frac{1}{2}, \frac{1}{2}, 1)$$

and its normalization $g'_2 = \frac{\sqrt{6}}{3}(-\frac{1}{2}, \frac{1}{2}, 1) = (-\frac{\sqrt{6}}{6}, \frac{\sqrt{6}}{6}, \frac{\sqrt{6}}{3})$. The pair $G' = (g'_1, g'_2)$ is an orthonormal basis for $V(G)$.

6.5 The QR Decomposition

The orthonormalization procedure described in the previous section also has an important consequence for the matrices we have been discussing. In fact, consider the matrix $M = M_G^E$ which, by the hypothesis on G (i.e. that G consists of an s-tuple of linearly independent vectors) has rank s. In addition let's consider the matrix $Q = M_{G'}^E$ and observe that it is an orthonormal matrix by construction. What is the relationship between M and Q? We already know that $M = M_G^E$, $Q = M_{G'}^E$ and thus $M = Q\, M_G^{G'}$.

If we look carefully at the formula $g'_s = \text{vers}\left(g_s - (g_s \cdot g'_1)g'_1 - \cdots - (g_s \cdot g'_{s-1})g'_{s-1}\right)$, we see that every vector g'_s of G' is a linear combination of vectors in G having subscript less than or equal to s. For example, $g'_2 = \text{vers}\left(g_2 - (g_2 \cdot g'_1)g_1\right)$ is a linear combination of g_2 and g_1. It's enough to reflect on this for a moment in order to understand immediately that the matrix $M_{G'}^G$ is upper triangular. Moreover, on the diagonal there are the reciprocals of the moduli of the non-zero vectors of G and hence are positive numbers . The matrix $M_G^{G'}$, which is the inverse of $M_{G'}^G$, has the same properties (see formula (1) in Section 3.5). In conclusion, putting $R = M_G^{G'}$, we can say that the matrix M can be put in the form

$$M = QR$$

which, in fact, is known as the QR **decomposition** or the QR **form** of M which encapsulates, in matrix form, the Gram-Schmidt procedure. We arrive at the following proposition.

Every matrix M of type (n, s) and rank s can be written in the form QR, where Q is orthonormal and R is upper triangular with positive diagonal.

Let's not get too nervous about the fact that we have to calculate the inverse of a matrix. In fact, we only have to find the inverse of a triangular matrix, and this operation is particularly easy. The reasons why this is easy follow from some considerations that we made in Section 3.3. We'll use them in the following example.

Example 6.5.1. We'll return to example 6.4.3. We had the matrix

$$M = \begin{pmatrix} 1 & 1 \\ 1 & 2 \\ 0 & 1 \end{pmatrix}$$

Recall that $g'_1 = \frac{1}{\sqrt{2}}g_1$ and hence $g_1 = \sqrt{2}g'_1$. Moreover, $g'_2 = \frac{\sqrt{6}}{3}\left(g_2 - \frac{3\sqrt{2}}{2}g'_1\right)$ and hence $g_2 = \frac{3\sqrt{2}}{2}g'_1 + \frac{\sqrt{6}}{2}g'_2$. Thus, $M = QR$, where

$$Q = \begin{pmatrix} \frac{\sqrt{2}}{2} & -\frac{\sqrt{6}}{6} \\ \frac{\sqrt{2}}{2} & \frac{\sqrt{6}}{6} \\ 0 & \frac{\sqrt{6}}{3} \end{pmatrix} \qquad R = \begin{pmatrix} \sqrt{2} & \frac{3\sqrt{2}}{2} \\ 0 & \frac{\sqrt{6}}{2} \end{pmatrix}$$

The method we just used above is not the only one you can use to arrive at the QR decomposition. Inasmuch as the result is so important, mathematicians have thrown themselves into the search for other ways to do this. So, to conclude this section in a worthy fashion let's look at an unexpected **alternate approach to calculating the QR decomposition** which uses the Cholesky decomposition.

We start off then, as before, with an s-tuple of linearly independent vectors $G = (g_1, \ldots, g_s)$ and consider the matrix $A = M_G^E$.

We already know that $A^{tr}A$ is positive semidefinite and now we will see that it is, in fact, positive definite. Recall that to prove this, it is enough to show that $AM_v^E = 0$ implies that $M_v^E = 0$, and this last thing is equivalent to the fact that the vectors of G are linearly independent. Thus we have verified that $A^{tr}A$ is positive definite and hence we know that it has a Cholesky decomposition. Thus we have

$$A^{tr}A = U^{tr}U$$

with U upper triangular and having positive diagonal. If we put $V = U^{-1}$ we also have that V is upper triangular with positive diagonal (see Section 3.5) and we get

$$V^{tr}A^{tr}AV = I$$

Putting $Q = AV$, the preceding formula says that $Q^{tr}Q = I$ and hence Q is orthonormal. Therefore

$$A = QU$$

which is the desired decomposition.

Exercises

Exercise 1. Consider the following matrix

$$A = \begin{pmatrix} 1 & 2 & 3 \\ 4 & 5 & 6 \\ 7 & 8 & 9 \end{pmatrix}$$

(a) Calculate the rank of A in two different ways: first as the maximum number of linearly independent rows and then as the maximum number of linearly independent columns.

(b) Calculate $B = A^T A$ and decide if B has a Cholesky decomposition.

Exercise 2. Consider the set S of all the matrices we get from I_3 by permuting its columns in all possible ways.

(a) How many matrices are there in S? (possible answers: 12, 3, 6, 4)

(b) Are they all orthogonal?

(c) Is the product of two matrices in S back again in S?

Exercise 3. Let G be the pair of vectors $((1,1,0),(-1,1,1))$ in $\mathbb{R}^3$ and let $V = V(G)$ be the space they generate.

(a) Decide if G is an orthonormal basis for V.

(b) Find all the vectors of $\mathbb{R}^3$ which have the same orthogonal projection on V as $(1,1,1)$.

(c) Setting $A = M_G^E$, find the QR decomposition of A.

Exercise 4. Consider the set E of all the orthonormal matrices in $\mathrm{Mat}_3(\mathbb{R})$ which have $(0\ 1\ 0)^{\mathrm{tr}}$ as their first column.

(a) Write down three distinct matrices in E.

(b) Are there symmetric matrices in E?

(c) Are there matrices in E with determinant 1?

Exercise 5. Let $v_1 = (1,0,3)$, $v_2 = (2,1,0)$, $v_3 = (3,1,3)$, $v = (3,3,3)$ be vectors in $\mathbb{R}^3$ and let $G = (v_1, v_2, v_3)$.

(a) Find the orthogonal projection, $p_{V(G)}(v)$, of v on $V(G)$.

(b) Find a vector $u \in \mathbb{R}^3$, $u \neq v$ such that $p_{V(G)}(v) = p_{V(G)}(u)$.

(c) Is the difference $u - v$ orthogonal to the vector $(1,1,-3)$?

Exercise 6. Consider the vectors $e_1 = (1,0,0)$, $e_3 = (0,0,1)$, $u_a = (1,a,2)$ in $\mathbb{R}^3$. Find the values of $a \in \mathbb{R}$ for which the orthogonal projection of e_1 and of e_3 on $V(u_a)$ coincide.

Exercise 7. We saw in this chapter: if $g_1, \ldots, g_s \in \mathbb{R}^n$ are such that $G = (g_1, \ldots, g_s)$ is an orthonormal basis for $V(G)$ and if $v \in \mathbb{R}^n$, then the orthogonal projection of v on $V(G)$ is given by the formula $p_{V(G)}(v) = (v \cdot g_1)g_1 + (v \cdot g_2)g_2 + \cdots + (v \cdot g_s)g_s$.
Prove that if G is only an orthogonal basis of $V(G)$ then one has

$$p_{V(G)}(v) = \frac{1}{|g_2|^2}(v \cdot g_1)g_1 + \frac{1}{|g_2|^2}(v \cdot g_2)g_2 + \cdots + \frac{1}{|g_s|^2}(v \cdot g_s)g_s$$

Exercise 8. Let n, r be natural numbers. Set I equal to the identity matrix of type n and let $Q \in \mathrm{Mat}_{n,r}(\mathbb{R})$ be an orthogonal matrix. Prove the following facts about the matrix $A = I - 2QQ^{\mathrm{tr}}$.

(a) A is symmetric.
(b) A is orthogonal.
(c) $A^2 = I$.

Exercise 9. Consider the three properties of the preceeding exercise and prove that any two of them implies the third.

Exercise 10. Find the QR decomposition of the following matrix

$$A = \begin{pmatrix} 2 & 3 & 4 & 2 & 2 \\ 0 & 1 & 7 & 1 & 6 \\ 0 & 0 & 10 & 4 & 6 \\ 0 & 0 & 0 & 1 & 3 \\ 0 & 0 & 0 & 0 & 3 \end{pmatrix}$$

Exercise 11. Consider the following system of homogeneous linear equations with coefficients in $\mathbb{R}$

$$\begin{cases} x_1 - x_2 + x_3 - 4x_4 - 4x_5 = 0 \\ x_1 - 5x_2 + x_3 - 14x_4 - 11x_5 = 0 \end{cases}$$

and let V be the vector subspace of $\mathbb{R}^5$ consisting of its solutions.

(a) Calculate a basis for V.
(b) Find the dimension of V.

ⓐ **Exercise 12.** Consider the following system of homogeneous linear equations with coefficients in $\mathbb{R}$

$$\begin{cases} 3x_1 - x_2 + x_3 - 4x_4 - 4x_5 = 0 \\ 2x_1 - 5x_2 + 2x_3 - 14x_4 - 11x_5 = 0 \\ -5x_1 + 3x_2 + x_3 - 7x_4 + 8x_5 = 0 \end{cases}$$

and let V be the vector subspace of $\mathbb{R}^5$ consisting of its solutions.

(a) Calculate a basis for V.
(b) Find the dimension of V.

7

Projections, Pseudoinverses and Least Squares

i cristalli sono lenti a volte immobili
e il vetro è spesso ma non sempre;
l'idea è sottile ma non c'entra

(From "RIFLESSIONI DI FRANCESCO"
by Francesco)

The farmers would say that we are fast approaching the harvest season. We wouldn't want to disillusion them and so we will begin (in this section) to harvest some of the fruit generated by the work of the previous sections. In particular we will discover how to solve, in various ways, a very important problem which goes by the name of *the problem of least squares*. But, we have to first equip ourselves with some important tools, namely *linear transformations*.

In mathematics we discover very quickly that right next to the idea of a linear object is the necessary idea of a linear transformation, sometimes referred to as a *homomorphism of vector spaces* (a strange name that actually sounds a bit threatening!). I would like to say that we are in for a surprise. But, in reality, it is not going to be a surprise to anyone if I say that the information of a linear transformation can be put in a matrix. We will begin with the idea of a *projection*. These are special matrices which allow us to extend the idea of orthogonal projection to spaces which are much more abstract than the physical spaces we are used to.

And what about matrices which don't have inverses? No fear, in this section we will also introduce the idea of a *pseudoinverse*. But, why have I even raised that question? And why this strange name? Don't lose heart, these are the tools we will use to deal with the problem of least squares. The reader will also soon discover what we mean by the phrase "least squares". We are not dealing with teeny weeny squares...

Robbiano L.: Linear Algebra for everyone
© Springer-Verlag Italia 2011

7.1 Matrices and Linear Transformations

We are about to rethink the discussion about projections of subspaces that we had in the preceding chapter (when we implemented the Gram-Schmidt procedure) and try to improve the consequences. Before doing that we have to make an important digression. We saw in Section 6.4 that, given an s-tuple of vectors $G = (v_1, \ldots, v_s)$ which are an orthonormal basis of $V(G)$, we can associate to every vector v in $\mathbb{R}^n$ the vector $p_{V(G)}(v)$, which is the orthogonal projection of v onto $V(G)$. When we do this operation for every vector we have defined a function

$$p_{V(G)} : \mathbb{R}^n \longrightarrow \mathbb{R}^n$$

What kind of function is this? In order to respond to this question let's do a few experiments. In particular let's take another look at Example 6.4.2 and let's calculate the three vectors $p_{V(G)}(e_1), p_{V(G)}(e_2), p_{V(G)}(e_3)$. We have

$$p_{V(G)}(e_1) = \tfrac{1}{|g_1|^2}(e_1 \cdot g_1)g_1 + \tfrac{1}{|g_2|^2}(e_1 \cdot g_2)g_2 = \tfrac{1}{2}(1, -1, 0) + \tfrac{1}{3}(1, 1, 1)$$
$$= (\tfrac{1}{2}, -\tfrac{1}{2}, 0) + (\tfrac{1}{3}, \tfrac{1}{3}, \tfrac{1}{3}) = (\tfrac{5}{6}, -\tfrac{1}{6}, \tfrac{1}{3})$$

$$p_{V(G)}(e_2) = \tfrac{1}{|g_1|^2}(e_2 \cdot g_1)g_1 + \tfrac{1}{|g_2|^2}(e_2 \cdot g_2)g_2 = -\tfrac{1}{2}(1, -1, 0) + \tfrac{1}{3}(1, 1, 1)$$
$$= (-\tfrac{1}{2}, \tfrac{1}{2}, 0) + (\tfrac{1}{3}, \tfrac{1}{3}, \tfrac{1}{3}) = (-\tfrac{1}{6}, \tfrac{5}{6}, \tfrac{1}{3})$$

$$p_{V(G)}(e_3) = \tfrac{1}{|g_1|^2}(e_3 \cdot g_1)g_1 + \tfrac{1}{|g_2|^2}(e_3 \cdot g_2)g_2 = 0 + \tfrac{1}{3}(1, 1, 1)$$
$$= (\tfrac{1}{3}, \tfrac{1}{3}, \tfrac{1}{3})$$

Recall that in Example 6.4.2 we calculated the projection of the vector $v = (2, 1, -7)$ onto the vector subspace $V = V(G)$ of $\mathbb{R}^3$. Notice the equality $v = 2e_1 + e_2 - 7e_3$ and let's calculate $2p_{V(G)}(e_1) + p_{V(G)}(e_2) - 7p_{V(G)}(e_3)$, i.e. the linear combination of the projections of the three vectors of the canonical basis made with the *same coefficients* with which we expressed v as a linear combination of the canonical basis. We obtain

$$2p_{V(G)}(e_1) + p_{V(G)}(e_2) - 7p_{V(G)}(e_3) = 2(\tfrac{5}{6}, -\tfrac{1}{6}, \tfrac{1}{3}) + (-\tfrac{1}{6}, \tfrac{5}{6}, \tfrac{1}{3}) - 7(\tfrac{1}{3}, \tfrac{1}{3}, \tfrac{1}{3})$$
$$= (-\tfrac{5}{6}, -\tfrac{11}{6}, -\tfrac{4}{3})$$

This is exactly the vector $p_{V(G)}(v)$ we already calculated in Example 6.4.2! Let's sum up what we just saw. We have before us an example for which the following property is true: if $v = a_1 e_1 + a_2 e_2 + a_3 e_3$, then its projection onto $V(G)$ is

$$p_{V(G)}(v) = a_1 p_{V(G)}(e_1) + a_2 p_{V(G)}(e_2) + a_3 p_{V(G)}(e_3)$$

In fact, if we look carefully at how we defined $p_{V(G)}$, it's clear that such a property is true not only for the vector v above, but for all the vectors of $\mathbb{R}^3$. And with just a bit more work we observe that the phenomenon we just saw is actually a property of all the functions of the type $p_{V(G)}$. Mathematicians say that the functions $p_{V(G)}$ *respect (don't change) linear combinations*.

By now the reader knows a bit about how things go in mathematics. Consequently, it won't surprise you when I ask the following question: are there other functions which *respect* linear combinations? If that question didn't leap spontaneously to your mind, don't worry; mathematicians have already answered this question. The answer is a result of extreme importance and soon we will see why. For the moment it is a good idea for us to experiment with some easy exercises to discover other functions with *this property*. An easy example is the identity function from $\mathbb{R}^n$ to $\mathbb{R}^n$. Another example of such a function is the function which associates to each vector its negative. A function which *does not have this property* is the function from $\mathbb{R}^2$ to $\mathbb{R}^2$ which associates to the vector (a_1, a_2) the vector (a_1^2, a_2^2). We see that $(2,2)$ is transformed into $(4,4)$ while we have $(2,2) = 2e_1 + 2e_2$ and yet e_1 is transformed into e_1, and e_2 is transformed into e_2. So, if the property were valid for this function we would have the equality $2e_1 + 2e_2 = (4,4)$, which is clearly not the case.

In fact, the property which is common to all the functions above (apart from the last one) can be described in the following way. Let φ be a function: for every relation between vectors of the type $v = a_1v_1 + \cdots, a_rv_r$ one has $\varphi(v) = a_1\varphi(v_1) + \cdots + a_r\varphi(v_r)$. Functions with this property are at the center of mathematics and, in particular, of linear algebra and are called **linear transformations**.

We have now arrived at another turning point. The following reasoning brings forward, for the n-th time, the central importance of the idea of a matrix and indeed, it cannot be said too often, provides the prerequisites for many of the most important applications of mathematics. Let φ be a linear transformation from $\mathbb{R}^c$ to $\mathbb{R}^r$. We will write

$$\varphi : \mathbb{R}^c \longrightarrow \mathbb{R}^r$$

Let $F = (f_1, \ldots, f_c)$ be a basis for $\mathbb{R}^c$ and $G = (g_1, \ldots, g_r)$ a basis for $\mathbb{R}^r$. If we know the vectors $\varphi(f_1), \ldots, \varphi(f_c)$, we can write them in a unique way as a linear combination of the elements of G. If we call $\varphi(F)$ the c-tuple $\varphi(f_1), \ldots, \varphi(f_c)$, then we know the matrix $M_{\varphi(F)}^G$. We can now see a very important fact. When we fix the two bases F and G, all the information about φ is contained in this matrix. In fact, if v is any vector whatsoever of $\mathbb{R}^c$ one has the equality, $v = FM_v^F$, where M_v^F is uniquely determined (recall that F is a basis). The linearity of φ implies that

$$\varphi(v) = \varphi(F)\, M_v^F \tag{1}$$

On the other hand we have just called $M_{\varphi(F)}^G$ the matrix for which

$$\varphi(F) = G\, M_{\varphi(F)}^G \tag{2}$$

Combining (1) and (2) one gets

$$\varphi(v) = G \, M^G_{\varphi(F)} \, M^F_v \tag{3}$$

This makes very clear the fact that, **having fixed the bases F and G, the information contained in φ is concentrated in the matrix $M^G_{\varphi(F)}$.** This observation confers the greatest importance on linear transformations. Moreover, since it is also true that

$$\varphi(v) = G \, M^G_{\varphi(v)} \tag{4}$$

we deduce the following formula

$$M^G_{\varphi(v)} = M^G_{\varphi(F)} \, M^F_v \tag{5}$$

If, instead of a single vector we had an s-tuple of vectors $S = (v_1, \ldots, v_s)$, we could apply (5) to all the vectors of S and obtain the following fundamental formula

$$M^G_{\varphi(S)} = M^G_{\varphi(F)} \, M^F_S \tag{6}$$

We note that if $\varphi : \mathbb{R}^c \longrightarrow \mathbb{R}^r$ is a linear transformation and if the two bases chosen are the canonical basis of $\mathbb{R}^c$ and of $\mathbb{R}^r$ (respectively) then formula (5) gives the following fact.

> If v is any vector in $\mathbb{R}^c$ then the components of v transformed by φ are linear homogeneous expressions in the components of v.

The last part of this section is dedicated to some mathematical ideas connected to linear transformations and is related to what was said at the end of Section 6.3.

Let φ be a linear transformation from $\mathbb{R}^c$ to $\mathbb{R}^r$. As we have already seen, the matrix $M^{E_r}_{\varphi(E_c)}$ contains all the information about φ. If we consider the set $\mathrm{Im}(\varphi)$, called the **image** of φ, i.e. the set of all the vectors in $\mathbb{R}^r$ which are the transforms of vectors in $\mathbb{R}^c$, we have the following rules, which an attentive reader should have no trouble in proving.

(1) The set $\mathrm{Im}(\varphi)$ is a vector subspace of $\mathbb{R}^r$.

(2) Let $t = \mathrm{rk}(M^{E_r}_{\varphi(E_c)})$ and choose t linearly independent columns of $M^{E_r}_{\varphi(E_c)}$, then the corresponding vectors form a basis for $\mathrm{Im}(\varphi)$.

(3) $\dim(\mathrm{Im}(\varphi)) = t$

If we denote by $\mathrm{Ker}(\varphi)$, called the **kernel** of φ, the set of all the vectors v in $\mathbb{R}^c$ for which $\varphi(v) = 0$, we have the following rules, which the reader can prove using rules (1) and (2) at the end of Section 6.3.

(1) The set $\mathrm{Ker}(\varphi)$ is a vector subspace of $\mathbb{R}^c$.

(2) Consider the system of linear equations $M_{\varphi(E_c)}^{E_r}\mathbf{x} = 0$. Then let $t = \mathrm{rk}(M_{\varphi(E_c)}^{E_r})$ and choose $c - t$ free variables and attribute to them the values $(1, 0, \ldots, 0)$, $(0, 1, \ldots, 0)$, $\ldots$, $(0, 0, \ldots, 1)$. The resulting solutions we obtain to the system of homogeneous linear equations form a basis for $\mathrm{Ker}(\varphi)$.

(3) $\dim(\mathrm{Ker}(\varphi)) = c - t$

Let's look at an example that will illustrate the new mathematical ideas which have just been introduced.

Example 7.1.1. Consider the function $\varphi : \mathbb{R}^3 \longrightarrow \mathbb{R}^2$ defined by the formula $\varphi(a, b, c) = (a + b,\ b - 2c)$. Notice that the components of $\varphi(v)$ are linear homogeneous expressions in the components of v and thus φ is a linear transformation. In order to calculate $M_{\varphi(E_3)}^{E_2}$ we must calculate the transformations of the members of the canonical basis of $\mathbb{R}^3$ and express those images using the canonical basis for $\mathbb{R}^2$. One has the equalities $\varphi(e_1) = (1, 0)$, $\varphi(e_2) = (1, 1)$, $\varphi(e_3) = (0, -2)$, from which it follows that the matrix $M_{\varphi(E_3)}^{E_2}$ is the following

$$M_{\varphi(E_3)}^{E_2} = \begin{pmatrix} 1 & 1 & 0 \\ 0 & 1 & -2 \end{pmatrix}$$

Now, let's do a little experiment. If v is the vector $(2, -2, 7)$, then applying the definition one has

$$\varphi(v) = (0, -16)$$

Applying formula (3) one gets

$$\varphi(v) = E_2 \begin{pmatrix} 1 & 1 & 0 \\ 0 & 1 & -2 \end{pmatrix} \begin{pmatrix} 2 \\ -2 \\ 7 \end{pmatrix} = 0e_1 - 16e_2 = (0, -16)$$

At this point it would be good if the reader expressed surprise at the coincidence. Let's try to verify the rules we described above. One sees that the matrix $M_{\varphi(E_3)}^{E_2}$ has rank 2, for example by observing that the first two columns are linearly independent. Thus, one has that $\dim(\mathrm{Im}(\varphi)) = 2$ and a basis for $\mathrm{Im}(\varphi)$ is, for example, $(\varphi(e_1), \varphi(e_2))$.

The homogeneous linear system associated to the matrix $M_{\varphi(E_3)}^{E_2}$ is the following

$$\begin{cases} x_1 + x_2 & = 0 \\ x_2 - 2x_3 = 0 \end{cases}$$

which transforms into the following

$$\begin{cases} x_1 & + 2x_3 = 0 \\ x_2 - 2x_3 = 0 \end{cases}$$

Let's consider x_3 as the free variable, and give it the value 1 and so obtain the solution $(-2, 2, 1)$. This vector is a basis for $\text{Ker}(\varphi)$ and, in fact, we know that all the solutions of the system are $(-2a,\ 2a,\ a)$, i.e. all multiples of the solution $(-2, 2, 1)$.

7.2 Projections

è difficile mettere a fuoco,
se brucia il proiettore

After the short excursion into the land of linear transformations that we made in the last section, let's return to the example that we used to motivate that excursion, i.e. orthogonal projections (see Section 6.4). Let's recall the nature of the problem. Suppose that we have a subspace V of $\mathbb{R}^n$ of dimension s and an orthonormal basis for it, say $G = (g_1, \ldots, g_s)$. So, $V = V(G)$ and $Q = M_G^E$ is an orthonormal matrix. At this point we begin a kind of technical reasoning that is typical of mathematics. I could avoid this and write only the final formula, but in this case, given the practical importance of the problem, an importance that goes beyond the theory, I have preferred to put myself in *math mode* and furnish a complete description of the steps that will bring us to the important conclusion. You'll also be able to see a mathematician at work. It is completely clear (really) that for every vector $v \in \mathbb{R}^n$ one has the equality $v = (e_1 \cdot v,\ e_2 \cdot v, \ldots, e_n \cdot v)$. We thus obtain the equality

$$
Q = \begin{pmatrix}
e_1 \cdot g_1 & e_1 \cdot g_2 & \cdots & e_1 \cdot g_s \\
e_2 \cdot g_1 & e_2 \cdot g_2 & \cdots & e_2 \cdot g_s \\
\vdots & \vdots & \vdots & \vdots \\
e_n \cdot g_1 & e_n \cdot g_2 & \cdots & e_n \cdot g_s
\end{pmatrix} \tag{1}
$$

To the s-tuple G we can associate the function $p_V : \mathbb{R}^n \longrightarrow \mathbb{R}^n$ defined by

$$
p_V(v) = (v \cdot g_1)g_1 + (v \cdot g_2)g_2 + \cdots + (v \cdot g_s)g_s \tag{2}
$$

which associates to every vector of $\mathbb{R}^n$ its orthogonal projection onto V. For simplicity we will call p this function p_V and observe that p is a linear transformation from $\mathbb{R}^n$ to $\mathbb{R}^n$. Choosing E as a basis for $\mathbb{R}^n$, the transformation is completely described by the matrix $M_{p(E)}^E$. Now we pose an important question, or better, we pose a question that will turn out to be important because of the importance of the answer. The question is: how is the matrix $M_{p(E)}^E$ made? From formula (2) we know that

$$
\begin{aligned}
p(e_1) &= (e_1 \cdot g_1)g_1 + (e_1 \cdot g_2)g_2 + \cdots + (e_1 \cdot g_s)g_s \\
p(e_2) &= (e_2 \cdot g_1)g_1 + (e_2 \cdot g_2)g_2 + \cdots + (e_2 \cdot g_s)g_s \\
\cdots &= \cdots\cdots \\
p(e_n) &= (e_n \cdot g_1)g_1 + (e_n \cdot g_2)g_2 + \cdots + (e_n \cdot g_s)g_s
\end{aligned} \tag{3}
$$

Moreover we observe that

$$Q^{\text{tr}} = \begin{pmatrix} e_1 \cdot g_1 & e_2 \cdot g_1 & \cdots & e_n \cdot g_1 \\ e_1 \cdot g_2 & e_2 \cdot g_2 & \cdots & e_n \cdot g_2 \\ \vdots & \vdots & \vdots & \vdots \\ e_1 \cdot g_s & e_2 \cdot g_s & \cdots & e_n \cdot g_s \end{pmatrix} \tag{4}$$

and thus (3) can be read as

$$p(E) = G\,Q^{\text{tr}} \tag{5}$$

from which we can immediately deduce the equality

$$M^G_{p(E)} = Q^{\text{tr}} \tag{6}$$

Using (3) and a *handy generalization* (mathematicians will forgive this expression, non-mathematicians won't worry about it) of formula (e) in Section 4.8, one can say that

$$M^E_{p(E)} = M^E_G\,M^G_{p(E)} \tag{7}$$

All that is left is to remember the equality $M^E_G = Q$ and expression (6) in order to reread (7) in the following way

$$M^E_{p(E)} = Q\,Q^{\text{tr}} \tag{8}$$

This is the first important answer. There still remains open another possibility which is the reason for the following question. *What happens if the basis for V is not orthonormal?* So, let's consider a non-orthonormal basis for V composed of an s-tuple which we will call $G' = (g'_1, \dots, g'_s)$. Put $M = M^E_{G'}$ and let's proceed by analogy with what was said in the special case where G' was orthonormal. First let's make a QR decomposition of M with Q orthonormal and R upper triangular with positive diagonal (see Section 6.5). We have

$$M = QR \quad \text{and so} \quad Q = MR^{-1} \tag{9}$$

Naturally, the columns of Q are formed by the coordinates of vectors in an orthonormal basis G of V and we have the equality

$$M = M^E_{G'} \qquad Q = M^E_G \qquad R = M^G_{G'} \tag{10}$$

Using (6), (10) and the identity $M^{G'}_{p(E)} = M^{G'}_G\,M^G_{p(E)}$ we obtain

$$M^{G'}_{p(E)} = R^{-1}Q^{\text{tr}} \tag{11}$$

and hence, multiplying by the identity matrix, $I = (R^{\text{tr}})^{-1}R^{\text{tr}}$, we have

$$M^{G'}_{p(E)} = R^{-1}(R^{\text{tr}})^{-1}R^{\text{tr}}Q^{\text{tr}} \tag{12}$$

From (12) and (9) we deduce that

$$M_{p(E)}^{G'} = R^{-1}(R^{\text{tr}})^{-1}M^{\text{tr}} \tag{13}$$

We also know that $Q^{\text{tr}}Q = I$ and from (9) it follows that

$$M^{tr}M = (QR)^{\text{tr}}QR = R^{\text{tr}}Q^{\text{tr}}QR = R^{\text{tr}}R \tag{14}$$

and hence we have

$$(M^{tr}M)^{-1} = (R^{\text{tr}}R)^{-1} = R^{-1}(R^{\text{tr}})^{-1} \tag{15}$$

Substituting into (13) we obtain a first important formula,

$$M_{p(E)}^{G'} = (M^{\text{tr}}M)^{-1}M^{\text{tr}} \tag{16}$$

A second important formula is obtained by combining (16) with the equality $M_{p(E)}^{E} = M_{G'}^{E} M_{p(E)}^{G'}$ in analogy with (7). We obtain

$$M_{p(E)}^{E} = M(M^{\text{tr}}M)^{-1}M^{\text{tr}} \tag{17}$$

Having arrived at the answer it's time to take a moment to catch our breath and review the situation.

We began with a subspace V of $\mathbb{R}^n$. We also were given a basis G' of V and we put $M = M_{G'}^{E}$ then

$$M_{p(E)}^{G'} = (M^{\text{tr}}M)^{-1}M^{\text{tr}} \qquad M_{p(E)}^{E} = M(M^{\text{tr}}M)^{-1}M^{\text{tr}} \qquad \text{(16) (17)}$$

Given an orthonormal basis G of V and putting $Q = M_{G}^{E}$, we have

$$M_{p(E)}^{G} = Q^{\text{tr}} \qquad M_{p(E)}^{E} = QQ^{\text{tr}} \qquad \text{(6) (8)}$$

Formulas (8) and (17) suggest that we consider a particular matrix to represent the orthogonal projection on the space generated by the columns of M. In fact, if M is a matrix of rank s in $\text{Mat}_{n,s}(\mathbb{R})$, the matrix $M_{p(E)}^{E} = M(M^{\text{tr}}M)^{-1}M^{\text{tr}}$ is called the **projection** onto the space generated by the columns of M, or more conveniently, the **projection onto** M. If M is orthonormal, the formula for the projection simplifies since we have the equality $M^{\text{tr}}M = I$, and then we have $M_{p(E)}^{G} = M^{\text{tr}}$, $M_{p(E)}^{E} = MM^{\text{tr}}$. In this form we see that formulas (6), (8) are special cases of formulas (16), (17). They seem very different only because the orthonormal matrices are usually called Q rather than M.

Example 7.2.1. Consider the vectors $v_1 = (1,0,1,0)$, $v_2 = (-1,-1,1,1)$ in $\mathbb{R}^4$. Let $F = (v_1, v_2)$ and let V be the subspace of $\mathbb{R}^4$ generated by F. The two vectors are linearly independent and hence F is a basis for V, but it is not an orthonormal basis. Thus, the projection onto V is obtained using formula (17). Setting $M = M_F^E$, we have

$$
M \, (M^{\text{tr}}M)^{-1}M^{\text{tr}} =
\begin{pmatrix}
1 & -1 \\
0 & -1 \\
1 & 1 \\
0 & 1
\end{pmatrix}
\left(
\begin{pmatrix}
1 & -1 \\
0 & -1 \\
1 & 1 \\
0 & 1
\end{pmatrix}^{\text{tr}}
\begin{pmatrix}
1 & -1 \\
0 & -1 \\
1 & 1 \\
0 & 1
\end{pmatrix}
\right)^{-1}
\begin{pmatrix}
1 & -1 \\
0 & -1 \\
1 & 1 \\
0 & 1
\end{pmatrix}^{\text{tr}}
$$

and making the calculations we see that

$$
M \, (M^{\text{tr}}M)^{-1}M^{\text{tr}} =
\begin{pmatrix}
\frac{3}{4} & \frac{1}{4} & \frac{1}{4} & -\frac{1}{4} \\
\frac{1}{4} & \frac{1}{4} & -\frac{1}{4} & -\frac{1}{4} \\
\frac{1}{4} & -\frac{1}{4} & \frac{3}{4} & \frac{1}{4} \\
-\frac{1}{4} & -\frac{1}{4} & \frac{1}{4} & \frac{1}{4}
\end{pmatrix}
$$

We conclude this section with two *mathematical gems*. We have just seen that if $M \in \text{Mat}_{n,s}(\mathbb{R})$ has rank s, the matrix $M(M^{\text{tr}}M)^{-1}M^{\text{tr}}$ is called the projection onto the space generated by the columns of M. What happens if the matrix doesn't have maximal rank?

Let's suppose that $A \in \text{Mat}_{n,r}(\mathbb{R})$ has rank $s < r$. From Section 6.3 we know that there are s columns of A that are linearly independent. If we call M the submatrix of A formed by those s columns, we have M of maximal rank and the vector space V generated by the columns of A coinciding with that generated by the columns of M. Thus, the projection onto V is $M(M^{\text{tr}}M)^{-1}M^{\text{tr}}$. The attentive reader will have noticed that the choice of the s linearly independent columns is not canonical. What happens if we make a different choice? Or, more generally, if we change the basis for V? And now we have the very satisfying result that **the projection doesn't depend on the basis chosen**. To some, perhaps, this not only seems a very satisfactory discovery but they would even like to see a proof. Let's try to please those people.

Let $A \in \text{Mat}_{n,r}(\mathbb{R})$ and let $s = \text{rk}(A)$. Let V be the vector space generated by the columns of A and let G be a basis for V formed by s linearly independent columns of A and let F be any other basis for V. Setting $M = M_G^E$, we have seen that the projection onto V is $M(M^{\text{tr}}M)^{-1}M^{\text{tr}}$. Setting $N = M_F^E$, and setting $P = M_F^G$ we have $M_F^E = M_G^E M_F^G$ and hence $N = MP$. In order to prove the independence which we spoke of above, we must establish the equality

$$
N(N^{\text{tr}}N)^{-1}N^{\text{tr}} = M(M^{\text{tr}}M)^{-1}M^{\text{tr}}
$$

Here's the proof.

$$
\begin{aligned}
N(N^{\mathrm{tr}}N)^{-1}N^{\mathrm{tr}} &= (MP)\big((MP)^{\mathrm{tr}}(MP)\big)^{-1}(MP)^{\mathrm{tr}} \\
&= (MP)\big((P^{\mathrm{tr}}M^{\mathrm{tr}})(MP)\big)^{-1}(MP)^{\mathrm{tr}} \\
&= (MP)\big(P^{\mathrm{tr}}(M^{\mathrm{tr}}M)P\big)^{-1}(MP)^{\mathrm{tr}} \\
&= (MP)\big(P^{-1}(M^{\mathrm{tr}}M)^{-1}(P^{\mathrm{tr}})^{-1}\big)(MP)^{\mathrm{tr}} \\
&= MPP^{-1}(M^{\mathrm{tr}}M)^{-1}(P^{\mathrm{tr}})^{-1}P^{\mathrm{tr}}M^{\mathrm{tr}} \\
&= M(M^{\mathrm{tr}}M)^{-1}M^{\mathrm{tr}}
\end{aligned}
$$

End of the proof.

The second gem is the following: **projections are symmetric matrix which are idempotent** i.e. matrices which equal their square **and positive semidefinite**. Would you like to see the proof? I am going to assume the answer is YES.

> In order to prove the matrix is symmetric it's enough to check that it is equal to its transpose (and that is easy). To prove the matrix is idempotent it's enough to calculate $M\,(M^{\mathrm{tr}}M)^{-1}M^{\mathrm{tr}}M\,(M^{\mathrm{tr}}M)^{-1}M^{\mathrm{tr}}$. Making the obvious simplifications we find the matrix with which we began. Now, calling such a matrix A we now know that $A = A^{\mathrm{tr}}$ and that $A = A^2$ and hence $A = AA = A^{\mathrm{tr}}A$, from which we conclude that A is positive semidefinite.

7.3 Least Squares and Pseudoinverses

In the last section we collected a bunch of mathematical facts. We are now adequately prepared and we can confront and resolve, with relative ease, the famous problem of least squares.

Problem of least squares: Case I . *If you are given an orthonormal matrix $Q \in \mathrm{Mat}_{n,s}(\mathbb{R})$ and a vector v in $\mathbb{R}^n$, how do you find a vector u which is both a linear combination of the columns of Q and at the same time has its distance from v a minimum?*

Solution. Let's call G the s-tuple of vectors whose coordinates, with respect to E are the columns of Q, i.e. $Q = M_G^E$. Given that we are looking for a vector of the form $u = GM_u^G$, we need to find M_u^G. We have already seen that the vector in $V(G)$ which is at a minimal distance from u is the vector $u = p(v)$ where $p = p_{V(G)}$. From formula (6) of Section 7.2 we deduce that

$$
M_{p(v)}^G = M_{p(E)}^G\, M_v^E = Q^{\mathrm{tr}}\, M_v^E
$$

The vector solution is thus

$$p(v) = G \, Q^{\mathrm{tr}} M_v^E \tag{18}$$

Given that $G = EM_G^E = EQ$, we also have

$$p(v) = E \, Q \, Q^{\mathrm{tr}} M_v^E \tag{19}$$

Problem of least squares: Case II. *If you are given a vector v in $\mathbb{R}^n$ and a matrix $M \in \mathrm{Mat}_{n,s}(\mathbb{R})$ of rank s, how do you find the vector which is both a linear combination of the columns of M and, at the same time, has minimum distance from v?*

Solution. As in the first case, let's call G' the s-tuple of vectors whose coordinates, with respect to E, are the columns of M, i.e. such that $M = M_{G'}^E$. Since we want $u = G'M_u^{G'}$, the problem asks us to find $M_u^{G'}$. We already said that $u = p(v)$ and from formula (16) one deduces that

$$M_{p(v)}^{G'} = M_{p(E)}^{G'} \, M_v^E = (M^{\mathrm{tr}}M)^{-1}M^{\mathrm{tr}} \, M_v^E$$

The solution vector is thus

$$p(v) = G' \, (M^{\mathrm{tr}}M)^{-1}M^{\mathrm{tr}}M_v^E \tag{20}$$

Given that $G' = EM_{G'}^E = EM$, one also has the equality

$$p(v) = E \, M \, (M^{\mathrm{tr}}M)^{-1}M^{\mathrm{tr}}M_v^E \tag{21}$$

Now is a good time to see an example.

Example 7.3.1. Let's have another look at Example 7.2.1. In particular, let's consider the vectors $v_1 = (1,0,1,0)$, $v_2 = (-1,-1,1,1)$ of $\mathbb{R}^4$, the pair $F = (v_1, v_2)$ and the subspace V of $\mathbb{R}^4$ generated by F. Set $M = M_F^E$. We calculated the projection $A = M \, (M^{\mathrm{tr}}M)^{-1}M^{\mathrm{tr}}$ and we obtained

$$A = \begin{pmatrix} \frac{3}{4} & \frac{1}{4} & \frac{1}{4} & -\frac{1}{4} \\ \frac{1}{4} & \frac{1}{4} & -\frac{1}{4} & -\frac{1}{4} \\ \frac{1}{4} & -\frac{1}{4} & \frac{3}{4} & \frac{1}{4} \\ -\frac{1}{4} & -\frac{1}{4} & \frac{1}{4} & \frac{1}{4} \end{pmatrix}$$

Now let's use A to calculate the orthogonal projection on the subspace V of an arbitrary vector and in this way solve the least squares problem. For example, if we let $e_1 = (1,0,0,0)$, then the vector in V at minimum distance from e_1 is the vector we get from formula (21). That vector is thus

$$E \, AM_{e_1}^E = (\frac{3}{4}, \frac{1}{4}, \frac{1}{4}, -\frac{1}{4})$$

In Section 6.4 we saw that the orthogonal projection on a vector subspace V can be calculated starting with any orthonormal basis of the subspace V. In Section 7.2 we saw that we could make that calculation starting with any basis of V, not necessarily even orthogonal. Some natural questions arise.

- Is it possible to calculate the orthogonal projection onto a subspace V starting only with a system of generators of V?
- Is it possible to solve the least squares problem for V starting with any system of generators for V?

We actually already know how to answer these questions. In fact, given a system of generators S of V and calling A the matrix M_S^E, it's enough to calculate $s = \mathrm{rk}(A)$, and then to let M be a submatrix of A formed by s linearly independent columns of A and calculate the projection $M(M^{\mathrm{tr}}M)^{-1}M^{\mathrm{tr}}$ (see formula (17) and (21) of the previous section).

Although we have answered the questions asked, they can be interpreted in another way. We could ask ourselves if it is possible to find a formula which only uses the matrix A directly (i.e. doesn't first try to find the submatrix M)? To answer this question we need a decomposition which we can associate to any matrix whatever. First let's look at an example.

Example 7.3.2. Let A be the matrix

$$
A = \begin{pmatrix} 1 & 1 & 0 & 1 \\ 0 & 1 & -2 & 3 \\ 2 & 1 & 2 & -1 \\ 1 & 2 & -2 & 4 \\ 1 & 0 & 2 & -2 \end{pmatrix} \in \mathrm{Mat}_{5,4}(\mathbb{R})
$$

Using the rules we saw in Section 6.3 it is possible to see that $\mathrm{rk}(A) = 2$ and that, for example, the first two columns of A are linearly independent. In particular, this implies that the first two columns of A are a basis for the subspace generated by the columns of the matrix. It follows that the third and fourth column are linear combinations of the first two. In fact, solving the two following two systems of linear equations

$$
\begin{cases} x_1 + x_2 = 0 \\ \quad\; x_2 = -2 \\ 2x_1 + x_2 = 2 \\ x_1 + 2x_2 = -2 \\ x_1 \qquad = 2 \end{cases}
\qquad
\begin{cases} x_1 + x_2 = 1 \\ \quad\; x_2 = 3 \\ 2x_1 + x_2 = -1 \\ x_1 + 2x_2 = 4 \\ x_1 \qquad = -2 \end{cases}
$$

we find the solution $(2, -2)$ for the first and $(-2, 3)$ for the second. These mean that the third column is twice the first minus twice the second, while the fourth column is minus twice the first plus three times the second. It follows that the matrix A may be represented as the product of two matrices of maximal rank equal to 2. In fact, we have the following identity

$$
\begin{pmatrix}
1 & 1 & 0 & 1 \\
0 & 1 & -2 & 3 \\
2 & 1 & 2 & -1 \\
1 & 2 & -2 & 4 \\
1 & 0 & 2 & -2
\end{pmatrix}
=
\begin{pmatrix}
1 & 1 \\
0 & 1 \\
2 & 1 \\
1 & 2 \\
1 & 0
\end{pmatrix}
\begin{pmatrix}
1 & 0 & 2 & -2 \\
0 & 1 & -2 & 3
\end{pmatrix}
$$

Using the same type of reasoning as in this example, we can see that such a decomposition is valid for any matrix. We can say the following.

Given a number field K and a matrix $A \in \mathrm{Mat}_{n,r}(K)$ of rank s, we can find two matrices $M \in \mathrm{Mat}_{n,s}(K)$ and $N \in \mathrm{Mat}_{r,s}(K)$, both of rank s, for which we have the following relation $A = MN^{\mathrm{tr}}$. This way of writing A is called the MN^{tr} form (or decomposition) of A.

Mathematicians denote by A^+ (or, unfortunately, as often happens, with other symbols) the matrix

$$
A^+ = N(N^{\mathrm{tr}}N)^{-1}(M^{\mathrm{tr}}M)^{-1}M^{\mathrm{tr}}
$$

and they call it the **pseudoinverse** or the **Moore-Penrose inverse** of A.

In the special case when $s = r$, the decomposition of A is simply $A = AI^{\mathrm{tr}}$, and in that case $A^+ = (A^{\mathrm{tr}}A)^{-1}A^{\mathrm{tr}}$. If, in addition, A is orthonormal, then $(A^{\mathrm{tr}}A) = I$ and hence $A^+ = A^{\mathrm{tr}}$.

The alert reader will have noticed that these matrices have already made their appearance in formulas (6) and (16). In Example 7.3.2 we have

$$
A^+ =
\begin{pmatrix}
\frac{1}{13} & 0 & \frac{2}{13} & \frac{1}{13} & \frac{1}{13} \\
\frac{19}{312} & \frac{1}{48} & \frac{21}{208} & \frac{17}{208} & \frac{25}{624} \\
\frac{5}{156} & \frac{-1}{24} & \frac{11}{104} & \frac{-1}{104} & \frac{23}{312} \\
\frac{3}{104} & \frac{1}{16} & \frac{-1}{208} & \frac{19}{208} & \frac{-7}{208}
\end{pmatrix}
$$

There are many things one can say about the notion of the pseudoinverse, but we will limit ourselves to a few essentials. The importance of the pseudoinverse comes from its valuable characteristics. In particular, the following properties are noteworthy.

(1) The matrices AA^+ and A^+A are symmetric.
(2) We have the following equality, $AA^+A = A$.
(3) The following equality is also true, $A^+AA^+ = A^+$.
(4) If A is invertible, then $A^+ = A^{-1}$.
(5) The matrix AA^+ is the projection on the space generated by the columns of A.

The proofs of these facts are easy and the attentive reader would not have to work very hard to construct the proofs. As you might guess, properties (2), (3) and (4) justify the name of pseudoinverse for the matrix A^+. Property (5) allows us to respond to the first question we raised above, i.e. is it possible to calculate the orthogonal projection on a subspace V starting just from an arbitrary system of generators of V? The answer is clear, the projection is simply AA^+.

There is, for sure, some reader, somewhere, who will make the following objection. The expression AA^+ depends on A only in appearance inasmuch as you need to find a submatrix M of A, of maximal rank, in order to calculate A^+. So, you might as well have used the expression $M(M^{tr}M)^{-1}M^{tr}$ directly to find the projection! Such an objection is very pertinent but there is an elegant escape route. It's clear that the decomposition of A as $A = MN^{tr}$ is not unique since it depends on the choice of the linearly independent columns of A that we choose. Fortunately (that is the way mathematicians express themselves!) we can prove that **the pseudoinverse does not depend on which submatrix M we choose** but depends only on the matrix A. The attentive reader will have noticed that this assertion is contained in the first of the two mathematical gems that we proved at the end of Section 7.2. We can be still more precise and prove that A^+ is **the unique matrix which enjoys the properties** (1), (2), (3) listed earlier. Using this fact, mathematicians have discovered that there are other methods for calculating A^+ that do not need the preliminary calculation of the matrix M.

It's also easy now to answer the second question we asked above, namely: is it possible to solve the least squares problem starting with any system of generators of the space V? We will put the problem in the form we have already used in the other cases.

Problem of least squares: Case III. *If we have a matrix $A \in \mathrm{Mat}_{n,r}(\mathbb{R})$ and a vector v in $\mathbb{R}^n$ how do we find a vector u which is both a linear combination of the columns of A and has its distance from v a minimum?*

Solution.   By looking at formula (21) and property (5) we conclude that the vector solution is

$$p(v) = E\, AA^+ M_v^E \qquad (22)$$

Let's take another look at Example 7.3.2.

Example 7.3.3. In Example 7.3.2 we had the matrix

$$A = \begin{pmatrix} 1 & 1 & 0 & 1 \\ 0 & 1 & -2 & 3 \\ 2 & 1 & 2 & -1 \\ 1 & 2 & -2 & 4 \\ 1 & 0 & 2 & -2 \end{pmatrix}$$

and we calculated its MN^{tr} decomposition as

$$M = \begin{pmatrix} 1 & 1 \\ 0 & 1 \\ 2 & 1 \\ 1 & 2 \\ 1 & 0 \end{pmatrix} \qquad N = \begin{pmatrix} 1 & 0 \\ 0 & 1 \\ 2 & -2 \\ -2 & 3 \end{pmatrix}$$

After that we obtained the equality

$$A^+ = \begin{pmatrix} \frac{1}{13} & 0 & \frac{2}{13} & \frac{1}{13} & \frac{1}{13} \\ \frac{19}{312} & \frac{1}{48} & \frac{21}{208} & \frac{17}{208} & \frac{25}{624} \\ \frac{5}{156} & \frac{-1}{24} & \frac{11}{104} & \frac{-1}{104} & \frac{23}{312} \\ \frac{3}{104} & \frac{1}{16} & \frac{-1}{208} & \frac{19}{208} & \frac{-7}{208} \end{pmatrix}$$

We can thus calculate

$$AA^+ = \begin{pmatrix} \frac{1}{6} & \frac{1}{12} & \frac{1}{4} & \frac{1}{4} & \frac{1}{12} \\ \frac{1}{12} & \frac{7}{24} & \frac{-1}{8} & \frac{3}{8} & \frac{-5}{24} \\ \frac{1}{4} & \frac{-1}{8} & \frac{5}{8} & \frac{1}{8} & \frac{3}{8} \\ \frac{1}{4} & \frac{3}{8} & \frac{1}{8} & \frac{5}{8} & \frac{-1}{8} \\ \frac{1}{12} & \frac{-5}{24} & \frac{3}{8} & \frac{-1}{8} & \frac{7}{24} \end{pmatrix}$$

and verify that AA^+ coincides, as expected by property (5), with the matrix $M\,(M^{\mathrm{tr}}M)^{-1}M^{\mathrm{tr}}$.

To close this section with a flourish let's look at the problem of least squares in a slightly different way. As the sages (and also the photographers) say, changing your point of view can reveal totally new aspects of an object. Up to now we have treated the problem of least squares *from a geometric point of view* linking it to orthogonal projections on a subspace. Let's see what the purely algebraic point of view is.

Let's suppose that we have a system of linear equations

$$A\mathbf{x} = \mathbf{b} \qquad\qquad (*)$$

Now observe that if A is invertible then $A^+ = A^{-1}$, by property (4). Thus, $A^+\mathbf{b} = A^{-1}\mathbf{b}$ is the solution of the system $(*)$. What if A is not invertible? What can we say about the column vector A^+b?

A fundamental observation is the following: **to say that the system has a solution is the same thing as saying that b is in the vector subspace generated by the columns of** A.

Another important observation is that substituting $A^+\mathbf{b}$ for $\mathbf{x}$ on the left side of the expression gives $AA^+\mathbf{b}$, i.e. the orthogonal projection of $\mathbf{b}$ on the space generated by the columns of A.

If the system has solutions then **b** is in the subspace generated by the columns of A hence the orthogonal projection of **b** coincides with **b**. In other words one has $AA^+\mathbf{b} = \mathbf{b}$ and hence $A^+\mathbf{b}$ is a solution of the system.

In case the system does not have solutions the column vector $AA^+\mathbf{b}$ is the vector in the space generated by the columns of A that is closest to **b**. We can thus conclude that if $(*)$ has no solutions then $A^+\mathbf{b}$ is the **best approximation to a solution (which doesn't exist)!**

There is no solution but there almost is! It's like life, in which not being able to do better we have to be content with what we get.

if you cannot make real what's ideal,
idealize what's real

Exercises

Exercise 1. Consider the following functions.

$\alpha : \mathbb{R}^3 \longrightarrow \mathbb{R}^3$ defined by $\alpha(a\ b\ c) = (a\ \ b+2c\ \ c-1)$
$\beta : \mathbb{R}^3 \longrightarrow \mathbb{R}^3$ defined by $\beta(a\ b\ c) = (a\ \ b+2c\ \ c-b)$
$\gamma : \mathbb{R}^3 \longrightarrow \mathbb{R}^3$ defined by $\gamma(a\ b\ c) = (a\ \ b+2c\ \ bc)$

(a) Decide which of these functions is linear.
(b) Let $A = M^E_{\alpha(E)}$. Prove, by finding an example, that A does not determine α.

Exercise 2. Consider the linear function $\varphi : \mathbb{R}^3 \longrightarrow \mathbb{R}^2$ defined by $\varphi(a\ b\ c) = (a+b\ \ b+2c)$.

(a) Find a non-zero vector $v \in \mathbb{R}^3$ such that $\varphi(v) = 0$.
(b) Verify that the 3-tuple $F = (f_1\ f_2\ f_3)$ with $f_1 = (1\ 0\ 1)$, $f_2 = (0\ 1\ -1)$, $f_3 = (3\ 3\ -5)$ is a basis for $\mathbb{R}^3$ and that the 2-tuple $G = (g_1\ g_2)$ with $g_1 = (2\ 1)$, $g_2 = (1\ -5)$ is a basis for $\mathbb{R}^2$.
(c) Calculate $M^G_{\varphi(F)}$.

Exercise 3. Consider the linear function $\varphi : \mathbb{R}^3 \longrightarrow \mathbb{R}^3$ defined by $\varphi(a\ b\ c) = (2a - 3b - c\ \ b + 2c\ \ 2a - 4b - 3c)$.

(a) Find a basis for $\mathrm{Im}(\varphi)$.
(b) Calculate $\dim(\mathrm{Im}(\varphi))$.
(c) Find a basis for $\mathrm{Ker}(\varphi)$.
(d) Calculate $\dim(\mathrm{Ker}(\varphi))$.

ⓐ **Exercise 4.** Consider the linear function $\varphi : \mathbb{R}^7 \longrightarrow \mathbb{R}^5$ defined by $\varphi(a_1\ a_2\ a_3\ a_4\ a_5\ a_6\ a_7) = (a_1 - a_6\ \ a_2 - a_7\ \ a_1 - a_4 - a_5 + a_6\ \ a_7\ \ a_3 - a_4)$.

(a) Find a basis for $\mathrm{Im}(\varphi)$.
(b) Calculate $\dim(\mathrm{Im}(\varphi))$.
(c) Find a basis for $\mathrm{Ker}(\varphi)$.
(d) Calculate $\dim(\mathrm{Ker}(\varphi))$.

Exercise 5. Consider the matrix

$$A = \begin{pmatrix} -4 & -1 \\ 0 & -1 \\ 1 & -1 \\ 1 & -1 \\ 3 & -1 \end{pmatrix} \in \mathrm{Mat}_{5\ 2}(\mathbb{R})$$

and calculate the projection onto A (i.e. onto the space spanned by the columns of A).

Exercise 6. Let $A = (2 \quad 1 \quad -2) \in \mathrm{Mat}_{1,3}(\mathbb{R})$.

(a) Calculate A^+.
(b) Calculate $(A^{\mathrm{tr}})^+$.

Exercise 7. Let $A = (a_1 \quad a_2 \quad \dots \quad a_n) \in \mathrm{Mat}_{1,n}(\mathbb{R})$. Put $c = \sum_{i=1}^{n} a_i^2$ and suppose that $c \neq 0$. Prove that $A^+ = \frac{1}{c} A^{\mathrm{tr}}$.

Exercise 8. Let n be a natural number and let u be a unit vector of $\mathbb{R}^n$. Let $A_u = M_u^E (M_u^E)^{\mathrm{tr}}$.

(a) Verify that $\mathrm{rk}(A_u) = 1$ for $u = \frac{1}{\sqrt{3}}(1,1,1)$ and for $u = \frac{1}{\sqrt{2}}(1,0,1)$.
(b) Prove that $\mathrm{rk}(A_u) = 1$ for every unit vector u.
(c) In case $n = 3$ give a geometric interpretation of (b), using projections.

Exercise 9. Let n, r be natural numbers. Set I equal to the identity matrix of type n, and let $Q \in \mathrm{Mat}_{n,r}(\mathbb{R})$ be an orthonormal matrix. Furthermore suppose that $A = I - 2QQ^{\mathrm{tr}}$. Prove the following assertions.

(a) The matrix A is symmetric.
(b) The matrix A is orthogonal.
(c) The matrix A satisfies, $A^2 = I$.

Exercise 10. Consider the vector $v = (3, 1, 3, 3) \in \mathbb{R}^4$, the matrix

$$A = \begin{pmatrix} 1 & 1 \\ 3 & -1 \\ 1 & 1 \\ 1 & 1 \end{pmatrix} \in \mathrm{Mat}_{4,2}(\mathbb{R})$$

and let V be the vector subspace of $\mathbb{R}^4$ generated by the columns of A.

(a) Prove that $v \in V$.
(b) Describe the set E of those vectors w having the following property: v is the vector in V having minimum distance from w.

Exercise 11. Let

$$A = \begin{pmatrix} -4 & 2 \\ 0 & 0 \\ 2 & -1 \\ 6 & -3 \\ 12 & -6 \\ 0 & 0 \end{pmatrix} \in \mathrm{Mat}_{6,2}(\mathbb{R})$$

(a) Set $s = \mathrm{rk}(A)$, and calculate two MN^{tr} decompositions of A with matrices $M, N \in \mathrm{Mat}_{6,s}(\mathbb{R})$ and $\mathrm{rk}(M) = \mathrm{rk}(N) = s$.
(b) Calculate A^+ in two ways, corresponding to the two decompositions you found above.

Exercise 12. This is a theoretical exercise.

(a) Let $M \in \mathrm{Mat}_{r,1}(\mathbb{R})$ and let $N \in \mathrm{Mat}_{1,r}(\mathbb{R})$. Prove that $\mathrm{rk}(MN) \leq 1$.

(b) Let $A, B \in \mathrm{Mat}_2(\mathbb{R})$. Prove that there exist matrices $C_1, C_2, C_3, C_4 \in \mathrm{Mat}_2(\mathbb{R})$ (of rank at most 1) for which we have the equality $AB = C_1 + C_2 + C_3 + C_4$.

(c) Deduce that if $A \in \mathrm{Mat}_2(\mathbb{R})$ is orthonormal then AA^{tr} is the sum of four projections of rank one.

@ Exercise 13. Consider the matrix A of Example 7.3.3 and verify properties c) and d), i.e. $AA^+A = A$ and $A^+AA^+ = A^+$.

@ Exercise 14. Consider Exercise 15 in Chapter 6.

(a) Using the pseudoinverse of A find a solution to $A\mathbf{x} = \mathbf{b}$ where

$$\mathbf{b} = (-62, 23, 163, 6, -106, -135)^{\mathrm{tr}}$$

(b) Using the pseudoinverse of A find an approximate solution of $A\mathbf{x} = b$ where $\mathbf{b} = (-62.01, 22.98, 163, 6, -106, -135)^{\mathrm{tr}}$.

@ Exercise 15. Let A be the matrix

$$A = \begin{pmatrix} -12 & -5 & -16 & 2 & 1 & -14 \\ 11 & -7 & -6 & 0 & 1 & 7 \\ 1 & 8 & 42 & -6 & -2 & 3 \\ 1 & -20 & -23 & 1 & 3 & -10 \\ 1 & -23 & -49 & 5 & 4 & -10 \\ 1 & 11 & -14 & 4 & -1 & 9 \end{pmatrix} \in \mathrm{Mat}_6(\mathbb{R})$$

(a) Find a basis for the vector subspace V of $\mathbb{R}^6$ generated by the columns of A.

(b) Calculate the pseudoinverse of A.

(c) Calculate the projection onto V in two different ways.

8

Endomorphisms and Diagonalization

> *the second thing I want to say*
> *is that I forgot the first thing I wanted to say*
> (from the volume I DON'T REMEMBER THE TITLE
> by an unknown author)

In this chapter we will study a central topic in the theory of matrices. We have already seen (and I hope we haven't forgotten) that matrices serve as containers of numerical information, for example as the fundamental parts of linear systems and as an essential mathematical tool for their solution. We've multiplied them, we've calculated their inverses (when they exist) and we have decomposed them into the LU decomposition. Then we began to appreciate them as geometric tools, in the sense that we have associated them to vectors and systems of coordinates.

Later we used them to describe and study quadratic forms, to construct orthonormal bases and to decompose them into the QR form. We've associated numerical invariants to them, for example the rank, and we saw they were useful in the study of vector subspaces of the space of n-tuples. Finally we used them extensively to study the problem of least squares.

But, to tell the truth, we have only fleetingly discussed one of the most important uses of matrices. We touched on this important use in Section 7.1 where matrices were used to help describe linear transformations. The mathematics used in that context was that of vector spaces and their transformations, even if (in fact) we didn't linger much on the technical subtleties.

In this, the final chapter of the book, we'll go into more depth about this last point and finally we will briefly see an extraordinary property of symmetric matrices with real entries. To put this last into perspective recall that every symmetric matrix is congruent to a diagonal matrix; this fact permitted us to find a basis so that the quadratic form associated with the symmetric matrix could be expressed (with respect to that basis) without any mixed terms.

Robbiano L.: Linear Algebra for everyone
© Springer-Verlag Italia 2011

Recall (see Section 5.2) that the relation of congruence is defined by the equation $B = P^{\mathrm{tr}} A P$ with P invertible. You may also recall that we emphasized the *extraordinary capacity* of matrices to adapt to very different situations. Because of this, it will probably not surprise you that we still have a few more surprises about matrices in store for you. In this chapter we will reveal them to you but first let's start by getting some idea of what I have in mind.

We already observed, at the end of the Section 7.1, that if $\varphi : \mathbb{R}^c \longrightarrow \mathbb{R}^r$ is a linear transformation, we have the formula

$$M^G_{\varphi(S)} = M^G_{\varphi(F)} \, M^F_S$$

It could happen that $c = r$. In that case, all the matrices in the game are square and F and G are bases *of the same space*. Thus, it makes sense to consider both $M^G_{\varphi(G)}$ and $M^F_{\varphi(F)}$. Are they related in any way? We'll see in a bit that they are and that the relation is not very different from that of congruence, even if the point of departure is completely different. In fact, the point of this chapter is to study this relationship, called *similarity*, and the chapter will conclude by showing that any real symmetric matrix is similar to a diagonal matrix. This result, as you will see, is *similar* to what we did for the relation of congruence, but the work needed to arrive at this result is much greater.

In fact, lots of matrices are similar to diagonal matrices. All we need to do is proceed in the opposite direction, i.e. take a diagonal matrix Δ, an invertible matrix P of the same type and construct the matrix $B = P^{-1}\Delta P$. A natural question to ask immediately is, what importance is there to the fact that B decomposes as $B = P^{-1}\Delta P$ with Δ diagonal. In order to begin to respond to this question, and also to motivate the reader who wants to understand this chapter (which is a bit more difficult than usual) let's perform an experiment.

Suppose that B is a square matrix of type 50 and suppose further that we want to calculate B^{100}. No problem, we begin by multiplying B with B and then multiply the result by B again, and so for 99 times. Each time we multiply by B the number of operations to do is on the order of $\frac{50^3}{3}$, as we saw in Section 2.3. So, the total number of operations to do is, on the order of, $99 * \frac{50^3}{3}$, i.e. *around 4 million calculations*. If, instead, Δ were a diagonal matrix of type 50, in order to calculate Δ^{100} it would be enough to raise each of the fifty diagonal element to the hundredth power, i.e. in total less than 5000 operations. And now the plot is revealed! If we knew a decomposition $B = P^{-1}\Delta P$, with Δ diagonal, then we would have the formula $B^{100} = P^{-1}\Delta^{100} P$. Now notice that multiplying a square matrix of type 50 by a diagonal matrix costs around 50^2 operations and multiplying two matrices of type 50 costs around $\frac{50^3}{3}$. Adding this all together we see that in this case we have a total of *around 50,000 operations*. That's not a bad saving!

Some attentive readers will have noticed that we pulled a bit of a fast one here in the sense that we are assuming that we know a decomposition of the form $B = P^{-1} \Delta P$. In fact, it is clear that to get such a decomposition we have to calculate it, if it exists, and that will cost as well. But, notice also that once we have done that work we can use the formula we had above to calculate *any power of B*.

So, we find ourselves in a situation analogous to the one we met when we discussed the *LU* decomposition (see Section 3.5) where, once the decomposition is calculated one can use it to solve, with less work, all the systems of linear equations which you get by varying the column of the constant terms.

The importance of these facts is truly extraordinary and, for example, will allow us to figure out how many rabbits there will be in a cage after a certain period of time (see Section 8.4)! If you don't really care much about rabbits, don't worry, there are plenty of other applications. Unfortunately, like most things in life, the grand conquests require a great deal of work. I suggest that the reader pay close attention to what follows.

8.1 An Example of a Plane Linear Transformation

Let's do a little numerical experiment. Consider the linear transformation $\varphi : \mathbb{R}^2 \longrightarrow \mathbb{R}^2$ such that $\varphi(e_1) = (\frac{5}{4}, \frac{\sqrt{3}}{4})$, $\varphi(e_2) = (\frac{\sqrt{3}}{4}, \frac{7}{4})$. One has

$$M_{\varphi(E)}^{E} = \begin{pmatrix} \frac{5}{4} & \frac{\sqrt{3}}{4} \\ \frac{\sqrt{3}}{4} & \frac{7}{4} \end{pmatrix} = \frac{1}{4} \begin{pmatrix} 5 & \sqrt{3} \\ \sqrt{3} & 7 \end{pmatrix}$$

Given that E is a basis for $\mathbb{R}^2$, from what we saw in Section 7.1 the matrix $M_{\varphi(E)}^{E}$ uniquely defines φ. Let $v_1 = (\sqrt{3}, -1)$, $v_2 = (1, \sqrt{3})$ and let $F = (v_1, v_2)$. We see that F is a basis for $\mathbb{R}^2$ and further that F has a noteworthy property. In fact

$$\varphi(v_1) = \frac{1}{4}(5\sqrt{3} - \sqrt{3}, \ \sqrt{3}\sqrt{3} - 7) = (\sqrt{3}, -1) = v_1$$

$$\varphi(v_2) = \frac{1}{4}(5 + \sqrt{3}\sqrt{3}, \ \sqrt{3} + 7\sqrt{3}) = (2, 2\sqrt{3}) = 2v_2$$

Summing up we see that we have found a basis of two vectors each of whose transform, by φ, has *not changed direction*. In fact, the vector v_1 is left *fixed*. If we represent this endomorphism φ using the basis F, i.e. if we write the matrix $M_{\varphi(F)}^{F}$, we get

$$M_{\varphi(F)}^{F} = \begin{pmatrix} 1 & 0 \\ 0 & 2 \end{pmatrix}$$

which is a *diagonal matrix*. This fact allows us to understand the geometric nature of the function φ and to describe it in the following way. Make a change of coordinates using F as the new basis (notice that F is orthogonal even if it is not orthonormal). With respect to the new set of axes given by F we observe that φ leaves unchanged the vectors of the new x'-axis and doubles the length of the vectors in the new y'-axis. In other words the function φ is a dilation along the direction of the new y'-axis, i.e. along the direction of the vector v_2.

Notice that we can also modify F in order to get an orthonormal basis F' by replacing v_1 and v_2 by their normalizations v'_1, v'_2. With respect to this new orthonormal basis F' the description of φ can be made with the same matrix, inasmuch as we have the relation

$$M^F_{\varphi(F)} = M^{F'}_{\varphi(F')} = \begin{pmatrix} 1 & 0 \\ 0 & 2 \end{pmatrix}$$

Given the nature of the construction, $M^E_{F'}$ is an orthonormal matrix and thus, by what was said in Section 6.2 the new coordinate system is obtained by a **rotation** of the old axes through some angle.

The new basis F' is a privileged basis and well adapted to reveal the true nature of the function φ. True enough, but where did the vectors v_1 and v_2 come from? Is there a way to calculate them, assuming that they exist no matter what the matrix $M^E_{\varphi(E)}$ is? We notice that the real substance of the discussion above was that the vectors v_1 and v_2 generate *privileged axes* for φ, thus our study should focus on the existence of *lines which are special for φ*. That is precisely what we will do in the next section.

8.2 Eigenvalues, Eigenvectors, Eigenspaces and Similarity

In this section we will introduce some rather *complex* mathematical concepts (the reason for the emphasis on the word above is clear to the specialists). For now, and also to complete the example in the previous section, let's make the following observation. Let $\varphi : \mathbb{R}^n \longrightarrow \mathbb{R}^n$ be a linear transformation (mathematicians often call them **endomorphisms of $\mathbb{R}^n$**). We can say that a basis F of $\mathbb{R}^n$ is special for φ if F is formed of vectors $v_1, \dots, v_n$ with the property that $\varphi(v_i)$ is a multiple of v_i for each $i = 1, \dots, n$. If $\varphi(v_i) = \lambda_i v_i$ for each $i = 1, \dots, n$ then we have

$$M^F_{\varphi(F)} = \begin{pmatrix} \lambda_1 & 0 & \cdots & 0 \\ 0 & \lambda_2 & \cdots & 0 \\ \vdots & \vdots & \vdots & \vdots \\ 0 & 0 & \cdots & \lambda_n \end{pmatrix}$$

a diagonal matrix. In order to find such a special basis the problem essentially becomes that of finding *non-zero* vectors v and *real numbers* λ such that

$\varphi(v) = \lambda v$. Such a number λ is called an **eigenvalue** of the endomorphism φ and the non-zero vector v is called an **eigenvector** of λ. All the vectors v for which $\varphi(v) = \lambda v$ form a vector subspace of $\mathbb{R}^n$, call the **eigenspace** of λ (*eigen* is German for *same*).

Let's stop for a moment and pick up on a discussion which we briefly mentioned in the introduction to this chapter, when we asked ourselves what was the relationship between the two matrices $M^G_{\varphi(G)}$, $M^F_{\varphi(F)}$. Let's suppose that we have two bases F, G of $\mathbb{R}^n$ (more generally of any vector space) and an endomorphism $\varphi : \mathbb{R}^n \longrightarrow \mathbb{R}^n$. We can thus form both $M^G_{\varphi(G)}$ and $M^F_{\varphi(F)}$ and it is not difficult to see the relationship between them: it's enough to apply rule (6) at the end of Section 7.1 and obtain $M^G_{\varphi(G)} = M^G_{\varphi(F)} M^F_G$. But we also have $M^G_{\varphi(F)} = M^G_F M^F_{\varphi(F)}$ by rule (e) of base change (see Section 4.8). So, in conclusion we have

$$M^G_{\varphi(G)} = M^G_F \, M^F_{\varphi(F)} \, M^F_G \tag{1}$$

Recall, from Section 4.7, that we have $M^G_F = (M^F_G)^{-1}$ and we now finally have the full picture on the relationship between $M^G_{\varphi(G)}$ and $M^F_{\varphi(F)}$. The two matrices are linked by a relation of the type

$$B = P^{-1} A \, P \tag{2}$$

that one calls the **relation of similarity**. Whenever A and B are two matrices for which formula (2) holds we say that B is **similar** to A. A closer look at this relationship shows that it is an equivalence relation, i.e. that A is similar to A, that if A is similar to B then also B is similar to A and finally that if A is similar to B and B is similar to C then A is similar to C.

One should say to the reader that these last comments are the sort of thing that mathematicians love to prove. In this case, however, the proofs are so easy that the readers should have no difficulty making them up.

But the interesting question is: what is the connection to eigenvalues and eigenvectors? We begin with a vitally important observation. Given a square matrix A of type n, we can think of the matrix as the matrix of an endomorphism φ. It's enough *to define* the endomorphism $\varphi : \mathbb{R}^n \longrightarrow \mathbb{R}^n$ using the formula $M^E_{\varphi(E)} = A$. This observation allows us to immediately speak of eigenvalues and eigenvectors not only of endomorphism but also of matrices. Now, changing the basis of $\mathbb{R}^n$ doesn't change the endomorphism but the matrix which represents the endomorphism does change according to formula (1) and hence gives a relation of type (2). As a consequence, associating eigenvalues and eigenvectors to matrices in the way we said above can only be justified if we can prove the following basic result, which is in fact true.

Similar matrices have the same eigenvalues and the same eigenvectors.

To arrive at an understanding of this fact we take two separate roads. The first uses some indirect reasoning, as follows. If A is a square matrix, we can put $A = M^E_{\varphi(E)}$, and in such a way define an endomorphism φ of $\mathbb{R}^n$ as we did above. If B is similar to A then $B = M^F_{\varphi(F)}$ where F is another basis of $\mathbb{R}^n$. Inasmuch as eigenvalues and eigenvectors are things that are intrinsic to φ and not to how φ is represented, we can conclude that similar matrices must have the same eigenvalues and eigenvectors.

The second approach confronts the question straight on. Using this approach we will obtain an even more important result. We proceed as follows. Given a square matrix A of type n consider the matrix $xI - A$ where I is the identity matrix of type n. The **characteristic polynomial of** A is the polynomial $p_A(x) = \det(xI - A)$. A moments reflection makes it clear that this is a polynomial in x of degree n. It is not a linear object, but linear algebra has a great need for this object and cannot do without it. Why is that? If λ is an eigenvalue of a given linear transformation $\varphi : \mathbb{R}^n \longrightarrow \mathbb{R}^n$ and we call $A = M^E_{\varphi(E)}$, then we know that $\lambda \in \mathbb{R}$ and that there is a non-zero vector v such that $\varphi(v) = \lambda v$; thus we deduce that M^E_v is a non-zero solution of the system of homogeneous linear equations $(\lambda I - A)\mathbf{x} = 0$. But, a homogeneous system of linear equations with as many equations as unknowns has a non-trivial solution if and only if the determinant of the coefficient matrix is zero.

It follows from this that the equality $p_A(\lambda) = 0$ with $\lambda \in \mathbb{R}$ is a necessary and sufficient condition in order for λ to be an eigenvalue of φ.

If B is a matrix which is similar to A then there is an invertible matrix P such that $B = P^{-1}AP$ (see formula (2)). It follows that

$$
\begin{aligned}
p_B(x) = \det(xI - B) &= \det(xI - P^{-1}AP) \qquad = \\
\det(P^{-1}xIP - P^{-1}AP) &= \det(P^{-1}(xI - A)P) \quad \text{and} \\
\det(P^{-1}(xI - A)P) &= \det(P^{-1})\det(xI - A)\det(P) = \\
\det(xI - A) &= p_A(x)
\end{aligned}
$$

Similar matrices have the same characteristic polynomial.

We have thus seen that the characteristic polynomial is, as mathematicians say, an invariant for similarity. This permits us to speak of **the characteristic polynomial of** φ. We have, in fact, that $p_\varphi(x) = p_A(x)$, where A is any matrix which represents φ.

The eigenvalues of φ (and of any matrix which represents it) are the real roots of the characteristic polynomial of φ (and of any matrix which represents it).

Having arrived at this point I can imagine that the reader might be a bit perplexed. One clearly sees that the reasoning in this chapter has been more difficult than that of the previous chapters. Perhaps at this point the book is not really for *everyone*? Don't despair, we are almost at the end and a small push is all that is needed to arrive at some really noteworthy conclusions. We still have to learn a couple of mathematical facts. The first is the following:

If for each eigenspace we select a basis and put all these bases in a single tuple of vectors we have a tuple of linearly independent vectors.

The second is the following:

The dimension of each eigenspace is less than or equal to the multiplicity of the root of the characteristic polynomial corresponding to that eigenspace.

These facts have some very important consequences but, fearing that the reader is beginning to lose patience, I will simply close this section with an example. The next section will show other classes of examples and will give ample witness to the versatility with which these mathematical facts can be used.

Example 8.2.1. Let's consider the following matrix $A = \left(\begin{smallmatrix} 7 & -6 \\ 8 & -7 \end{smallmatrix}\right)$. Its characteristic polynomial is

$$p_A(x) = \det \begin{pmatrix} x - 7 & 6 \\ -8 & x + 7 \end{pmatrix} = x^2 - 1$$

The eigenvalues are thus $1, -1$. We can now calculate the corresponding eigenspaces V_1, V_{-1}. In order to calculate V_1 we have to find all the vectors v such that $\varphi(v) = v$. But, what is the φ we are talking about? We thought of A as $M^E_{\varphi(E)}$ and therefore if $v = (x_1, x_2)^{\mathrm{tr}}$, then the vector v for which $\varphi(v) = v$ are those for which the coordinates, with respect to the canonical basis, are the solutions to the linear system

$$\begin{cases} -6x_1 + 6x_2 = 0 \\ -8x_1 + 8x_2 = 0 \end{cases}$$

A basis for V_1 is, for example, the vector $u_1 = (1, 1)$. As far as V_{-1} is concerned, the vectors v for which $\varphi(v) = -v$ are precisely those vectors whose coordinates, with respect to the canonical basis, are the solutions to the system of linear equations

$$\begin{cases} -8x_1 + 6x_2 = 0 \\ -8x_1 + 8x_2 = 0 \end{cases}$$

So, a basis for V_1 is, for example, the vector $u_1 = (3, 4)$. The conclusion of all this discussion is that we can form the matrix $P^{-1} = \left(\begin{smallmatrix} 1 & 3 \\ 1 & 4 \end{smallmatrix}\right)$, and then we have $P = \left(\begin{smallmatrix} 4 & -3 \\ -1 & 1 \end{smallmatrix}\right)$, and, setting $\Delta = \left(\begin{smallmatrix} 1 & 0 \\ 0 & -1 \end{smallmatrix}\right)$, we get the decomposition $A = P^{-1} \Delta P$, i.e.

$$\begin{pmatrix} 7 & -6 \\ 8 & -7 \end{pmatrix} = \begin{pmatrix} 1 & 3 \\ 1 & 4 \end{pmatrix} \begin{pmatrix} 1 & 0 \\ 0 & -1 \end{pmatrix} \begin{pmatrix} 4 & -3 \\ -1 & 1 \end{pmatrix}$$

8.3 Powers of Matrices

The power of matrices or the powers of matrices? The emphasis has always been on the first interpretation, but in this section we will be talking about the second, as we indicated in the introduction to this chapter. Without any delay let's look at an example.

Example 8.3.1. Let's consider the following matrix

$$A = \begin{pmatrix} 2 & 0 & -3 \\ 1 & 1 & -5 \\ 0 & 0 & -1 \end{pmatrix}$$

and let's suppose that we want to raise it to a very high power, for example, 50,000. As we observed in the introduction, it is quite costly to calculate $A^{50,000}$, but in this case we can take advantage of eigenvalues. How? The characteristic polynomial of A is

$$p_A(x) = \det(xI - A) = x^3 - 2x^2 - x + 2 = (x+1)(x-1)(x-2)$$

There are three distinct eigenvalues -1, 1, 2. Three corresponding eigenvectors are $v_1 = (1,2,1)$, $v_2 = (0,1,0)$, $v_3 = (1,1,0)$.
    Set

$$P^{-1} = \begin{pmatrix} 1 & 0 & 1 \\ 2 & 1 & 1 \\ 1 & 0 & 0 \end{pmatrix} \qquad \Delta = \begin{pmatrix} -1 & 0 & 0 \\ 0 & 1 & 0 \\ 0 & 0 & 2 \end{pmatrix}$$

and we get

$$P = \begin{pmatrix} 0 & 0 & 1 \\ -1 & 1 & -1 \\ 1 & 0 & -1 \end{pmatrix} \qquad A = P^{-1}\Delta P$$

As we saw in the introduction, we have the equality

$$A^N = P^{-1}\Delta^N P$$

Calculating Δ^N is an easy operation, in fact we have

$$\Delta^N = \begin{pmatrix} (-1)^N & 0 & 0 \\ 0 & 1^N & 1 \\ 0 & 0 & 2^N \end{pmatrix}$$

The upshot of all of this is the following formula

$$\begin{pmatrix} 2 & 0 & -3 \\ 1 & 1 & -5 \\ 0 & 0 & -1 \end{pmatrix}^N = \begin{pmatrix} 1 & 0 & 1 \\ 2 & 1 & 1 \\ 1 & 0 & 0 \end{pmatrix} \begin{pmatrix} (-1)^N & 0 & 0 \\ 0 & 1^N & 1 \\ 0 & 0 & 2^N \end{pmatrix} \begin{pmatrix} 0 & 0 & 1 \\ -1 & 1 & -1 \\ 1 & 0 & -1 \end{pmatrix}$$

Thus, if N is an even number we have

$$\begin{pmatrix} 2 & 0 & -3 \\ 1 & 1 & -5 \\ 0 & 0 & -1 \end{pmatrix}^N = \begin{pmatrix} 2^N & 0 & -2^N+1 \\ 2^N-1 & 1 & -2^N+1 \\ 0 & 0 & 1 \end{pmatrix}$$

If, instead, N is an odd number we have

$$\begin{pmatrix} 2 & 0 & -3 \\ 1 & 1 & -5 \\ 0 & 0 & -1 \end{pmatrix}^N = \begin{pmatrix} 2^N & 0 & -2^N+1 \\ 2^N-1 & 1 & -2^N-3 \\ 0 & 0 & -1 \end{pmatrix}$$

At this point the reader might be curious to verify if, indeed, calculating A^N with the obvious method, i.e. $AAAAAA \cdots A$ with $N-1$ products, is really trumped by the method we talked about using the diagonalization of A. Naturally, to really appreciate the difference in this *contest* we have to choose N big enough. How big? I prefer to leave that choice to the discretion of the reader.

8.4 The Rabbits of Fibonacci

> *moreover,*
> *those who truly wish to acquire some skill in this science,*
> *must apply themselves continually*
> *exercising with it, practically on a daily basis*
>
> (Leonardo da Pisa, called Fibonacci)

As G.K. Chesterton once observed, with persistence even the snail reached Noah's Ark. Thus, even though the quote above was written in 1202 in the LIBER ABACI of Fibonacci, it is still good advice today.

Leonardo da Pisa, also called *Bigollo* (the ne'er do well), who is also now known by the name of *filius Bonacci* (son of Bonacci) or Fibonacci, was a great mathematician who lived at the end of the twelfth and the beginning of the thirteenth centuries.

Ne'er do well and mathematician! Yet he is so important that even today his book, Liber Abaci, influences modern science. For example, we owe to him the use of the symbol 0, which he imported to the West from the great Indo-Arab tradition. Once, during a tourney, someone posed the following question to him.

> *In a cage, which has no opening to the outside world, there are a pair of rabbits. Let's suppose that rabbits give birth every month, starting the second month of their lives. Suppose also that every time they give birth they have one male and one female offspring. How many pairs of rabbits will there be in the cage after one year?*

Clearly this does not describe the behavior of *real rabbits*. But, one knows that mathematicians (even a ne'er do well) like to extract simplified mathematical models from real life situations. Perhaps after solving this problem one could complicate the model and have Fibonacci's rabbits behave more like real rabbits (who give birth in a much less regular way, have a certain mortality, don't live in ideal cages...).

Let's consider how to pose the problem. At the beginning, i.e. after zero months, there is one pair of rabbits (which we will call C) in the cage. After one month there will still be the same pair of rabbits. However, after two months there is the pair C plus the newly born offspring, which we will call F, hence now there are two pairs. After three months there is the pair C, the pair F and the new offspring of C, which we'll call G. After four months we have: the pair C, the pair F, the pair G, another new pair which are offspring of C, and a new pair which are the offspring of F, for a grand total of 5 pairs. Let's try to visualize this with the following table

months	0	1	2	3	4	⋯
couples	1	1	2	3	5	⋯

Now we have to figure out how to go on, but that problem is easily resolved. In fact, having made a thought experiment, it should be clear that after n months there are: the number of pairs there were after $n - 1$ months, plus the offspring of the pairs there were after $n - 2$ months, since every pair of rabbits gives (as offspring) precisely one new pair. If we call, for simplicity, $C(n)$, the number of pairs there are after n months, one has the following formula

$$C(n) = C(n-1) + C(n-2) \qquad \text{with} \quad C(1) = C(0) = 1 \qquad (1)$$

This type of formula is called a **linear recurrence** where the initial data are $C(1) = C(0) = 1$. Thus, it is easy to extend the table above.

months	0	1	2	3	4	5	6	7	8	9	10	11	12	⋯
couples	1	1	2	3	5	8	13	21	34	55	89	144	233	⋯

Thus, Fibonacci's problem (calculate $C(12)$) is now done. After a year there will be 233 pairs in the cage. The reader will have noticed that to calculate $C(12)$ we had to first calculate all the preceding values of $C(n)$ i.e. $C(0)$, $C(1), \ldots, C(11)$. Now, after several centuries, we have become more demanding and we would like to have a more direct formula, i.e. a formula that allows us to calculate $c(12)$ without calculating all the preceding values.

In a bit we will see that this is possible using eigenvalues. So, the reader should now pay close attention because we are arriving at the most important point of this example. The first thing to do is to find a matrix! Since, up to this moment no matrix has appeared in this problem it is not at all apparent how we can use the notion of an eigenvalue. But, let's look again at formula (1). It can also be described in the following way

$$\begin{aligned} C(n) &= C(n-1) + C(n-2) \\ C(n-1) &= C(n-1) \end{aligned} \qquad \text{with} \quad C(1) = C(0) = 1 \qquad (2)$$

In other words one has

$$\begin{pmatrix} C(n) \\ C(n-1) \end{pmatrix} = \begin{pmatrix} 1 & 1 \\ 1 & 0 \end{pmatrix} \begin{pmatrix} C(n-1) \\ C(n-2) \end{pmatrix} \qquad \text{with} \quad C(1) = C(0) = 1 \qquad (3)$$

As often happens in mathematics, the simple fact of adding a trivial equation, i.e. $C(n-1) = C(n-1)$, to the mix has completely changed our view of the problem and opened up the possibility of viewing it in a different way. Let's see how. If we apply formula (3) to the case $n = 2$ we get

$$\begin{pmatrix} C(2) \\ C(1) \end{pmatrix} = \begin{pmatrix} 1 & 1 \\ 1 & 0 \end{pmatrix} \begin{pmatrix} C(1) \\ C(0) \end{pmatrix} \qquad (4)$$

If we apply formula (3) to the case $n = 3$ we get

$$\begin{pmatrix} C(3) \\ C(2) \end{pmatrix} = \begin{pmatrix} 1 & 1 \\ 1 & 0 \end{pmatrix} \begin{pmatrix} C(2) \\ C(1) \end{pmatrix}$$

Using (4) one gets

$$\begin{pmatrix} C(3) \\ C(2) \end{pmatrix} = \begin{pmatrix} 1 & 1 \\ 1 & 0 \end{pmatrix} \begin{pmatrix} 1 & 1 \\ 1 & 0 \end{pmatrix} \begin{pmatrix} C(1) \\ C(0) \end{pmatrix} = \begin{pmatrix} 1 & 1 \\ 1 & 0 \end{pmatrix}^2 \begin{pmatrix} C(1) \\ C(0) \end{pmatrix} \qquad (5)$$

At this point it should be clear that, continuing in the same way we obtain

$$\begin{pmatrix} C(n) \\ C(n-1) \end{pmatrix} = \begin{pmatrix} 1 & 1 \\ 1 & 0 \end{pmatrix}^{n-1} \begin{pmatrix} C(1) \\ C(0) \end{pmatrix} \qquad (6)$$

for every $n \geq 2$. Notice that once again we are faced with a matrix raised to a power. Let's see now if we can calculate the eigenvalues of the matrix $A = \begin{pmatrix} 1 & 1 \\ 1 & 0 \end{pmatrix}$. It's characteristic polynomial is

$$p_A(x) = \det(xI - A) = \det \begin{pmatrix} x-1 & -1 \\ -1 & x \end{pmatrix} = x^2 - x - 1$$

The real roots of the characteristic polynomial, i.e. the eigenvalues of A, are

$$x_1 = \frac{1 + \sqrt{5}}{2}, \qquad x_2 = \frac{1 - \sqrt{5}}{2}$$

Now we can calculate the eigenspaces and, preceding as in Example 8.2.1, we get

$$\begin{pmatrix} 1 & 1 \\ 1 & 0 \end{pmatrix} = \begin{pmatrix} \frac{1+\sqrt{5}}{2} & \frac{1-\sqrt{5}}{2} \\ 1 & 1 \end{pmatrix} \begin{pmatrix} \frac{1+\sqrt{5}}{2} & 0 \\ 0 & \frac{1-\sqrt{5}}{2} \end{pmatrix} \begin{pmatrix} \frac{1}{\sqrt{5}} & \frac{-1+\sqrt{5}}{2\sqrt{5}} \\ \frac{-1}{\sqrt{5}} & \frac{1+\sqrt{5}}{2\sqrt{5}} \end{pmatrix} \qquad (7)$$

We have almost arrived at the central point. In the introduction to this chapter and also in the preceding section we have seen that the n-th power of a diagonal matrix is easy to calculate (it's enough to raise the diagonal entries to the n-th power) and also that if $A = P^{-1}\Delta P$ with Δ diagonal then $A^n = P^{-1}\Delta^n P$. Hence, from formulas (6), (7) and (3) we deduce

$$\begin{pmatrix} C(n) \\ C(n-1) \end{pmatrix} = \begin{pmatrix} \frac{1+\sqrt{5}}{2} & \frac{1-\sqrt{5}}{2} \\ 1 & 1 \end{pmatrix} \begin{pmatrix} \frac{1+\sqrt{5}}{2} & 0 \\ 0 & \frac{1-\sqrt{5}}{2} \end{pmatrix}^{n-1} \begin{pmatrix} \frac{1}{\sqrt{5}} & \frac{-1+\sqrt{5}}{2\sqrt{5}} \\ \frac{-1}{\sqrt{5}} & \frac{1+\sqrt{5}}{2\sqrt{5}} \end{pmatrix} \begin{pmatrix} 1 \\ 1 \end{pmatrix} \quad (1)$$

Now, using the preceding formula one finds

$$C(n) = \frac{1}{\sqrt{5}} \left(\left(\frac{1+\sqrt{5}}{2} \right)^{n+1} - \left(\frac{1-\sqrt{5}}{2} \right)^{n+1} \right) \quad (8)$$

We have finally arrived! Let's verify, for example, that $C(3) = 3$. According to formula (8) we have

$$C(3) = \frac{1}{\sqrt{5}} \left(\left(\frac{1+\sqrt{5}}{2} \right)^4 - \left(\frac{1-\sqrt{5}}{2} \right)^4 \right)$$
$$= \frac{1}{\sqrt{5}} \left(\left(\frac{1+4\sqrt{5}+6*25+20\sqrt{5}+25}{16} \right) - \left(\frac{1-4\sqrt{5}+6*25-20\sqrt{5}+25}{16} \right) \right)$$
$$= \frac{1}{\sqrt{5}} \left(\frac{48\sqrt{5}}{16} \right) = 3$$

8.5 Differential Systems

You've read that properly, we will be talking about systems of differential equations. It's true, even in analysis they use matrices! Let's look at an example right away.

Example 8.5.1. We begin by recalling that the differential equation

$$x'(t) = c\,x(t) \quad (1)$$

where $x(t)$ is a function of time t, and c is a constant, has as its solution

$$x(t) = x(0)\,e^{ct} \quad (2)$$

which takes into account the initial value $x(0)$. What would happen if, instead of having a scalar equation, we had a vector equation? Suppose we have modeled a problem, for example the relationship between populations of predators and prey, with the following system which makes very evident the two quantities and their derivatives at certain times.

$$\begin{cases} x_1'(t) = 2x_1(t) - 3x_2(t) \\ x_2'(t) = x_1(t) - 2x_2(t) \end{cases} \quad (3)$$

Let's simplify the notation a bit by not always writing t.

$$\begin{cases} x_1' = 2x_1 - 3x_2 \\ x_2' = x_1 - 2x_2 \end{cases} \qquad (4)$$

The idea is to read the system (4) as a matrix equation. It will be enough to consider the column vector $\mathbf{x} = (x_1, x_2)^{tr}$, the matrix $A = \begin{pmatrix} 2 & -3 \\ 1 & -2 \end{pmatrix}$ and hence to rewrite the system as

$$\mathbf{x}' = A\mathbf{x} \qquad (5)$$

Now, let's try to diagonalize the matrix A. It's characteristic polynomial is $\det(xI - A) = x^2 - 1$ and thus A has eigenvalues $1, -1$. Doing some simple calculations we see that an eigenvector for the eigenvalue 1 is $v_1 = (3, 1)$, while an eigenvector for the eigenvalue -1 is $v_2 = (1, 1)$. Using the usual relations $M^E_{\varphi(E)} = M^E_F M^F_{\varphi(F)} M^F_E$ and putting $\Delta = \begin{pmatrix} 1 & 0 \\ 0 & -1 \end{pmatrix}$, $P = M^F_E = \begin{pmatrix} \frac{1}{2} & -\frac{1}{2} \\ -\frac{1}{2} & \frac{3}{2} \end{pmatrix}$, we have $P^{-1} = M^E_F = \begin{pmatrix} 3 & 1 \\ 1 & 1 \end{pmatrix}$ and

$$A = P^{-1}\Delta P \qquad (6)$$

i.e.

$$\begin{pmatrix} 2 & -3 \\ 1 & -2 \end{pmatrix} = \begin{pmatrix} 3 & 1 \\ 1 & 1 \end{pmatrix} \begin{pmatrix} 1 & 0 \\ 0 & -1 \end{pmatrix} \begin{pmatrix} \frac{1}{2} & -\frac{1}{2} \\ -\frac{1}{2} & \frac{3}{2} \end{pmatrix} \qquad (7)$$

Now comes the good idea. Substitute in equality (5) the expression for A given in (6) and multiply on the left by P. We get

$$P\mathbf{x}' = \Delta P\mathbf{x} \qquad (8)$$

and setting

$$\mathbf{y} = P\mathbf{x} \qquad (9)$$

we obtain, because of the linearity of the derivative,

$$\mathbf{y}' = P\mathbf{x}' \qquad (10)$$

and hence (8) can be rewritten

$$\mathbf{y}' = \Delta\mathbf{y} \qquad (11)$$

i.e.

$$\begin{cases} y_1' = y_1 \\ y_2' = -y_2 \end{cases} \qquad (12)$$

All the work we have done up to this point allows us to transform the system (4) into system (12). But, what is the advantage of doing that? The attentive reader will certainly have noticed that in system (12) the **variables are separated** and thus the two equations can be solved separately, as was done for case (1). Thus we obtain (see (2))

$$\begin{cases} y_1(t) = y_1(0)\, e^t \\ y_2(t) = y_2(0)\, e^{-t} \end{cases} \qquad (13)$$

From (9) we have that $\mathbf{y}(0) = P\mathbf{x}(0)$ and hence

$$\begin{cases} y_1(0) = & \frac{1}{2}x_1(0) - \frac{1}{2}x_2(0) \\ y_2(0) = & -\frac{1}{2}x_1(0) + \frac{3}{2}x_2(0) \end{cases} \tag{14}$$

From (9) we also conclude that $\mathbf{x} = P^{-1}\mathbf{y}$ and using (13) we obtain

$$\begin{cases} x_1(t) = 3y_1(0)e^t + y_2(0)e^{-t} \\ x_2(t) = \ y_1(0)e^t + y_2(0)e^{-t} \end{cases} \tag{15}$$

To finish, it is enough to use (14).

$$\begin{cases} x_1(t) = 3\big(\frac{1}{2}x_1(0) - \frac{1}{2}x_2(0)\big)e^t + \big(-\frac{1}{2}x_1(0) + \frac{3}{2}x_2(0)\big)e^{-t} \\ x_2(t) = \ \big(\frac{1}{2}x_1(0) - \frac{1}{2}x_2(0)\big)e^t + \big(-\frac{1}{2}x_1(0) + \frac{3}{2}x_2(0)\big)e^{-t} \end{cases} \tag{16}$$

The easy verification that these functions actually satisfy the system (3) with which we began, will convince the reader that the result is correct.

I cannot forego the opportunity to insist once more that matrices are one of the most important tools in mathematics. At this stage of the game, the reader has seen their enormous versatility and should have no doubt about their value.

8.6 Diagonalizability of Real Symmetric Matrices

We have just about arrived at the most important result. But, once more we need to make a small digression, this time of a type we haven't seen in the past. Let's consider the following three polynomials $F_1(x) = x^2 - 2x - 1$, $F_2(x) = x^2 - x + 1$ e $F_3(x) = x^2 - 2x + 1$. The first has two real distinct roots $1 - \sqrt{2}$, $1 + \sqrt{2}$, while the second doesn't have real roots (it has two complex roots $\frac{1-\sqrt{3}\,i}{2}$, $\frac{1+\sqrt{3}\,i}{2}$). The third is a square, in fact $F_3(x) = (x-1)^2$ and hence has only one real root, but mathematicians prefer (and they have good reasons) to say that it has *two coincident roots*, or, better yet, *it has one root with multiplicity 2*. In general, a polynomial with real coefficients has complex roots that, counted with their multiplicity, are as many as the degree of the polynomial. This fact is so important that it is called, with a trumpet fanfare, **the fundamental theorem of algebra**. But, as we have already noted with the polynomial F_2, the polynomial may not have any real roots.

We are just about ready to see a really great new theorem, but first we would like to respond to a suspicion that surely some readers have posed in their minds. Could it be true that all real square matrices are similar to diagonal matrices, i.e. as one would say in the mathematical jargon, are they all **diagonalizable**? The answer is a resounding no and we immediately give an example.

Example 8.6.1. Consider the matrix $A = \left(\begin{smallmatrix} 1 & 1 \\ 0 & 1 \end{smallmatrix}\right)$. It's characteristic polynomial is the following

$$p_A(x) = \det(xI - A) = (x - 1)^2$$

This matrix has only one eigenvalue $\lambda_1 = 1$ with multiplicity equal to 2. Let's calculate the eigenspace V_1. So, we have to find all the vectors v such that $\varphi(v) = v$, where φ is the linear transformation on $\mathbb{R}^2$ into itself defined by $M^E_{\varphi(E)} = A$. Thus, we have to find the vectors v such that $AM^E_v = M^E_v$ i.e. such that $(A - I)M^E_v = 0$. In other words, we have to solve the linear homogeneous system of equations

$$\begin{cases} x_2 = 0 \\ 0 = 0 \end{cases} \tag{1}$$

The general solution is $(a, 0)$. A basis for V_i can be chosen to be the unit vector $(1, 0)$. We have arrived at a place beyond which we are unable to go, in the sense that the eigenvectors are too few to be able to construct a basis for $\mathbb{R}^2$ made up entirely of eigenvectors. The only possible conclusion is that we have found an example of a matrix which is not diagonalizable.

Notice that the matrix in the preceding example is not symmetric and the first noteworthy fact about diagonalization and symmetric matrices is the following.

The characteristic polynomial of a real symmetric matrix has all of its roots real.

This means that, if counted with their multiplicities, there are as many eigenvalues as the degree of the characteristic polynomial which, in turn, coincides with the type of the matrix A. But, you might say that this was also true for the characteristic polynomial of the previous example, and you would be correct. But, there is (in reserve) another fundamentally important fact.

The eigenspaces corresponding to different eigenvalues of a symmetric matrix are pairwise orthogonal.

This implies that it is possible to find **an orthonormal basis made entirely of eigenvectors**. Let's see a proof of this fact.

Suppose that λ_1, λ_2 are two distinct eigenvalues and suppose that u is a non-zero eigenvector of λ_1, v a non-zero eigenvector of λ_2. Put $x = M^E_u$, $y = M^E_v$. Then one has

$$\lambda_1(x^{\mathrm{tr}}\, y) = (\lambda_1 x)^{\mathrm{tr}}\, y = (Ax)^{\mathrm{tr}}\, y = x^{\mathrm{tr}}\, A^{\mathrm{tr}}\, y$$

$$= x^{\mathrm{tr}}\, Ay = x^{\mathrm{tr}}\, (Ay) = x^{\mathrm{tr}}\, (\lambda_2 y) = \lambda_2(x^{\mathrm{tr}}\, y)$$

The equality $\lambda_1(x^{\mathrm{tr}}\, y) = \lambda_2(x^{\mathrm{tr}}\, y)$ implies $x^{\mathrm{tr}}\, y = 0$ and, as a consequence implies $u \cdot v = 0$.

You might say, even in the example above the property just discussed was true, and you would be right, in fact, the eigenspace corresponding to the eigenvalue 0 is the null space and thus the property just announced is trivially true. But, in the end, the symmetric matrices deliver their final gift.

The eigenspaces corresponding to the eigenvalues of a symmetric matrix have dimension equal to their multiplicity.

This property was not satisfied in the preceding example! The strong, and surprising, consequence of all this is the following.

Real symmetric matrices are diagonalizable and the diagonalization can be done with an orthonormal matrix. In other words given a real symmetric matrix $A \in \mathrm{Mat}_n(\mathbb{R})$, there exists an orthonormal basis F of $\mathbb{R}^n$ and a diagonal matrix Δ such that

$$\Delta = P^{-1}A\,P \tag{2}$$

with Δ diagonal and $P = M_F^E$. **The matrix Δ has, on its main diagonal, the eigenvalues of A, each repeated as many times as its multiplicity.**

In order to be able to put this river of important mathematical facts into perspective, let's study (in detail) some examples.

Example 8.6.2. Let's look at the symmetric matrix

$$A = \begin{pmatrix} 1 & 1 & 0 \\ 1 & -2 & 3 \\ 0 & 3 & 1 \end{pmatrix}$$

Consider the matrix

$$xI - A = \begin{pmatrix} x-1 & -1 & 0 \\ -1 & x+2 & -3 \\ 0 & -3 & x-1 \end{pmatrix}$$

and its determinant which is the characteristic polynomial of A

$$p_A(x) = \det(xI - A) = x^3 - 13x + 12 = (x-3)(x-1)(x+4)$$

There are, as is easy to check, three distinct eigenvalues, $\lambda_1 = 3$, $\lambda_2 = 1$, $\lambda_3 = -4$. Let's calculate the three corresponding eigenspaces, which we will call V_1, V_2, V_3. To calculate V_1 we have to find all the vectors v such that $\varphi(v) = 3v$. But, what's the φ we are talking about? We mean the linear transformation φ of $\mathbb{R}^3$ into itself, such that $M_{\varphi(E)}^E = A$. So, we must find the vectors v such that $AM_v^E = 3M_v^E$ i.e. such that $(A-3I)M_v^E = 0$ or equivalently $(3I - A)M_v^E = 0$. In other words we have to solve the homogeneous system of linear equations

$$\begin{cases} 2x_1 - & x_2 & = 0 \\ -x_1 + & 5x_2 - 3x_3 = 0 \\ & -3x_2 + 2x_3 = 0 \end{cases} \tag{1}$$

One sees that the general solution to this system is $(a, 2a, 3a)$. An orthonormal basis of V_1 is thus given by the single vector of unit length, $f_1 = \frac{1}{\sqrt{14}}(1, 2, 3)$.

Repeating the same sort of reasoning for V_2 we see that we have to solve the following homogeneous system of linear equations

$$\begin{cases} & -x_2 & = 0 \\ -x_1 + & 3x_2 - 3x_3 = 0 \\ & -3x_2 & = 0 \end{cases} \tag{2}$$

The general solution of this system is $(3a, 0, -a)$. An orthonormal basis for V_2 is thus given by the single vector of unit length $f_2 = \frac{1}{\sqrt{10}}(3, 0, -1)$.

Repeating the same reasoning one more time for V_3 we are faced with the homogeneous system of linear equations

$$\begin{cases} -5x_1 - & x_2 & = 0 \\ -x_1 - & 2x_2 - 3x_3 = 0 \\ & -3x_2 - 5x_3 = 0 \end{cases} \tag{3}$$

Here the general solution of the system is $(a, -5a, 3a)$. An orthonormal basis for V_3 is thus given by the single vector of unit length which is $f_3 = \frac{1}{\sqrt{35}}(1, -5, 3)$.

Let's put together the orthonormal bases we found for V_1, V_2, V_3 to get the basis $F = (f_1, f_2, f_3)$ of $\mathbb{R}^3$. By the way it was constructed, we have

$$M_{\varphi(F)}^F = \begin{pmatrix} 3 & 0 & 0 \\ 0 & 1 & 0 \\ 0 & 0 & -4 \end{pmatrix}$$

The matrix

$$M_F^E = \begin{pmatrix} \frac{1}{\sqrt{14}} & \frac{3}{\sqrt{10}} & \frac{1}{\sqrt{35}} \\ \frac{2}{\sqrt{14}} & 0 & \frac{-5}{\sqrt{35}} \\ \frac{3}{\sqrt{14}} & \frac{-1}{\sqrt{10}} & \frac{3}{\sqrt{35}} \end{pmatrix}$$

is orthonormal, hence the inverse M_E^F is the transpose of M_F^E. The formula $M_{\varphi(F)}^F = M_E^F \, M_{\varphi(E)}^E \, M_F^E$ can be explicitly written down as

$$\begin{pmatrix} 3 & 0 & 0 \\ 0 & 1 & 0 \\ 0 & 0 & -4 \end{pmatrix} = \begin{pmatrix} \frac{1}{\sqrt{14}} & \frac{2}{\sqrt{14}} & \frac{3}{\sqrt{14}} \\ \frac{3}{\sqrt{10}} & 0 & \frac{-1}{\sqrt{10}} \\ \frac{1}{\sqrt{35}} & \frac{-5}{\sqrt{35}} & \frac{3}{\sqrt{35}} \end{pmatrix} \begin{pmatrix} 1 & 1 & 0 \\ 1 & -2 & 3 \\ 0 & 3 & 1 \end{pmatrix} \begin{pmatrix} \frac{1}{\sqrt{14}} & \frac{3}{\sqrt{10}} & \frac{1}{\sqrt{35}} \\ \frac{2}{\sqrt{14}} & 0 & \frac{-5}{\sqrt{35}} \\ \frac{3}{\sqrt{14}} & \frac{-1}{\sqrt{10}} & \frac{3}{\sqrt{35}} \end{pmatrix}$$

We have obtained the diagonalization of A. Setting $P = M_F^E$, and

$$\Delta = \begin{pmatrix} 3 & 0 & 0 \\ 0 & 1 & 0 \\ 0 & 0 & -4 \end{pmatrix}$$

we have

$$\Delta = P^{-1}A\,P = P^{\mathrm{tr}}\,A\,P$$

Notice that not only are Δ and A **similar** but they are also **congruent**.

Example 8.6.3. Let A be the symmetric matrix

$$A = \begin{pmatrix} \frac{9}{10} & -\frac{1}{5} & \frac{1}{2} \\ -\frac{1}{5} & \frac{3}{5} & 1 \\ \frac{1}{2} & 1 & -\frac{3}{2} \end{pmatrix}$$

Consider the matrix

$$xI - A = \begin{pmatrix} x - \frac{9}{10} & \frac{1}{5} & -\frac{1}{2} \\ \frac{1}{5} & x - \frac{3}{5} & -1 \\ -\frac{1}{2} & -1 & x + \frac{3}{2} \end{pmatrix}$$

and its determinant, which is the characteristic polynomial of A,

$$p_A(x) = \det(xI - A) = x^3 - 3x + 2 = (x-1)^2(x+2)$$

We have thus found two distinct eigenvalues $\lambda_1 = 1$, $\lambda_2 = -2$, but notice that λ_1 has multiplicity 2.

Let's calculate the corresponding eigenspaces, that we will call V_1, V_2. In order to calculate V_1 we have to find all the vectors v such that $\varphi(v) = v$. As in the preceding example, φ is the linear transformation of $\mathbb{R}^3$ into itself for which $M_{\varphi(E)}^E = A$. Thus we have to find the vectors v such that $AM_v^E = M_v^E$, i.e. such that $(A - I)M_v^E = 0$ or equivalently $(I - A)M_v^E = 0$. In other words we have to solve the homogeneous system of linear equations

$$\begin{cases} \frac{1}{10}x_1 + \frac{1}{5}x_2 - \frac{1}{2}x_3 = 0 \\ \frac{1}{5}x_1 + \frac{2}{5}x_2 - 1x_3 = 0 \\ -\frac{1}{2}x_1 - x_2 + \frac{5}{2}x_3 = 0 \end{cases} \tag{1}$$

The system is equivalent to its last equation and thus its general solution is $(a, -\frac{1}{2}a + \frac{5}{2}b, b)$. A basis of V_1 is thus given by the pair of vectors (v_1, v_2) where $v_1 = (1, -\frac{1}{2}, 0)$, $v_2 = (0, \frac{5}{2}, 1)$. If we want an orthonormal basis it will be enough to use the Gram-Schmidt method. In this way we obtain the new basis $G = (g_1, g_2)$, where $g_1 = \mathrm{vers}(v_1)$, $g_2 = \mathrm{vers}(v_2 - (v_2 \cdot g_1)g_1)$. Thus we have $g_1 = \frac{1}{\sqrt{5}}(2, -1, 0)$ and $g_2 = \mathrm{vers}\big((0, \frac{5}{2}, 1) + \frac{\sqrt{5}}{2}\frac{1}{\sqrt{5}}(2, -1, 0)\big) = \mathrm{vers}(1, 2, 1)$ and hence $g_2 = \frac{1}{\sqrt{6}}(1, 2, 1)$.

Repeating the same reasoning for V_2 we find that we have to solve the system of homogeneous linear equations

$$\begin{cases} -\frac{29}{10}x_1 + \frac{1}{5}x_2 - \frac{1}{2}x_3 = 0 \\ \frac{1}{5}x_1 + \frac{13}{5}x_2 - x_3 = 0 \\ -\frac{1}{2}x_1 - x_2 - \frac{1}{2}x_3 = 0 \end{cases} \tag{2}$$

The general solution for the system is $(-a, -2a, 5a)$. An orthonormal basis of V_2 is given by the single vector of unit length $g_3 = \frac{1}{\sqrt{30}}(-1, -2, 5)$.

Putting together the orthonormal bases of both V_1 and V_2 already found, we obtain the basis $F = (g_1, g_2, g_3)$ of $\mathbb{R}^3$. By the way this basis was constructed we have

$$M^F_{\varphi(F)} = \begin{pmatrix} 1 & 0 & 0 \\ 0 & 1 & 0 \\ 0 & 0 & -2 \end{pmatrix}$$

The matrix

$$M^E_F = \begin{pmatrix} \frac{2}{\sqrt{5}} & -\frac{1}{\sqrt{6}} & \frac{1}{\sqrt{30}} \\ -\frac{1}{\sqrt{5}} & \frac{2}{\sqrt{6}} & -\frac{2}{\sqrt{30}} \\ 0 & \frac{1}{\sqrt{6}} & \frac{5}{\sqrt{30}} \end{pmatrix}$$

is orthonormal and hence the inverse of M^F_E coincides with the transpose of M^E_F. The formula $M^F_{\varphi(F)} = M^F_E M^E_{\varphi(E)} M^E_F$ can be written explicitly as

$$\begin{pmatrix} 1 & 0 & 0 \\ 0 & 1 & 0 \\ 0 & 0 & -2 \end{pmatrix} = \begin{pmatrix} \frac{2}{\sqrt{5}} & -\frac{1}{\sqrt{5}} & 0 \\ -\frac{1}{\sqrt{6}} & \frac{2}{\sqrt{6}} & \frac{1}{\sqrt{6}} \\ \frac{1}{\sqrt{30}} & -\frac{2}{\sqrt{30}} & \frac{5}{\sqrt{30}} \end{pmatrix} \begin{pmatrix} \frac{9}{10} & -\frac{1}{5} & \frac{1}{2} \\ -\frac{1}{5} & \frac{3}{5} & 1 \\ \frac{1}{2} & 1 & -\frac{3}{2} \end{pmatrix} \begin{pmatrix} \frac{2}{\sqrt{5}} & -\frac{1}{\sqrt{6}} & \frac{1}{\sqrt{30}} \\ -\frac{1}{\sqrt{5}} & \frac{2}{\sqrt{6}} & -\frac{2}{\sqrt{30}} \\ 0 & \frac{1}{\sqrt{6}} & \frac{5}{\sqrt{30}} \end{pmatrix}$$

We have thus obtained the diagonalization of A. Writing $P = M^E_F$, and

$$\Delta = \begin{pmatrix} 1 & 0 & 0 \\ 0 & 1 & 0 \\ 0 & 0 & -2 \end{pmatrix}$$

we have $\Delta = P^{-1}A P = P^{\mathrm{tr}} A P$.

We finish with an interesting observation. If A is a real symmetric matrix then we have seen that there exists a diagonal matrix Δ and an orthonormal matrix P such that $\Delta = P^{-1}AP$. The matrix Δ has the eigenvalues of A on its main diagonal. The observation (already made at the end of Example 8.6.2) is that inasmuch as P is orthonormal, and consequently $P^{-1} = P^{\mathrm{tr}}$, A and Δ are not only similar they are also congruent. As a consequence of this observation we have the following fact.

If A is a real symmetric positive semidefinite matrix then its eigenvalues are all non-negative. If, moreover, A is positive definite then its eigenvalues are all positive.

As you might well imagine, mathematicians have studied the idea of diagonalization a great deal and made many important discoveries about it. But, having come this far the road ahead is much less accessible and to go ahead is no longer *for everyone*. Nevertheless, for the readers who have had their thirst for the subject only partially sated by what they have already learned, my advice, obviously, is: don't stop here!

$$e_{\ell_{a\ sete\ s}a^{\ell^e}}$$

(from PALINDROMES OF (LO)RENZO
from Lorenzo)

Exercises

Exercise 1. For every real number φ let A_φ be the following matrix

$$A_\varphi = \begin{pmatrix} \cos(\varphi) & -\sin(\varphi) \\ \sin(\varphi) & \cos(\varphi) \end{pmatrix}$$

(a) Prove that for every $\varphi \in \mathbb{R}$ there exists $\vartheta \in \mathbb{R}$ such that $(A_\varphi)^{-1} = A_\vartheta$.
(b) Describe all the values of $\varphi \in \mathbb{R}$ for which the matrix A_φ is diagonalizable.
(c) Give geometric motivations for the answers to the preceding questions.

Exercise 2. How many, and what are, the orthogonal and diagonalizable matrices in $\text{Mat}_2(\mathbb{R})$?

Exercise 3. Consider the following matrices.

$$A_1 = \begin{pmatrix} 1 & 0 \\ 0 & 1 \end{pmatrix} \quad A_2 = \begin{pmatrix} 1 & 0 \\ 1 & 1 \end{pmatrix} \quad A_3 = \begin{pmatrix} 1 & 0 \\ 1 & 2 \end{pmatrix} \quad A_4 = \begin{pmatrix} 2 & 0 \\ 1 & 1 \end{pmatrix} \quad A_5 = \begin{pmatrix} 2 & 0 \\ 0 & 1 \end{pmatrix}$$

(a) Explain which of these are diagonalizable.
(b) Find the pairs of matrices, among these, that are similar.

Exercise 4. Discuss the following theoretical questions.

(a) What are the eigenvalues of an upper triangular matrix?
(b) What are the eigenvalues of a lower triangular matrix?
(c) Prove that if λ is an eigenvalue of the matrix A and N is a natural number then λ^N is an eigenvalue of the matrix A^N.

Exercise 5. Solve the following system of differential equations

$$\begin{cases} x_1'(t) = x_1(t) - 3x_2(t) \\ x_2'(t) = -3x_1(t) + 10x_2(t) \end{cases}$$

with initial conditions $x_1(0) = 2$, $x_2(0) = -4$.

Exercise 6. Consider the matrix

$$A = \begin{pmatrix} 0 & 0 & -2 \\ 1 & 2 & 1 \\ 1 & 0 & 3 \end{pmatrix}$$

(a) Calculate $\det(A)$ and verify that the matrix A is invertible.
(b) Deduce that 0 is not an eigenvalue of A.
(c) Diagonalize, if possible, A.

@ Exercise 7. Diagonalize the following matrix

$$\begin{pmatrix} \frac{55010}{32097} & \frac{3907}{64194} & \frac{58286}{32097} & -\frac{42489}{21398} & \frac{61403}{64194} & -\frac{6067}{64194} \\ -\frac{65587}{32097} & \frac{128036}{32097} & \frac{809651}{32097} & -\frac{218715}{10699} & \frac{448180}{32097} & \frac{10561}{32097} \\ -\frac{3821}{32097} & -\frac{25687}{64194} & -\frac{71447}{32097} & \frac{10299}{21398} & -\frac{110219}{64194} & \frac{21601}{64194} \\ \frac{8672}{32097} & \frac{11408}{32097} & -\frac{83864}{32097} & \frac{16268}{10699} & -\frac{35260}{32097} & \frac{2336}{32097} \\ \frac{5507}{10699} & \frac{5014}{10699} & -\frac{51355}{10699} & \frac{57171}{10699} & -\frac{13254}{10699} & -\frac{6221}{10699} \\ -\frac{2818}{32097} & \frac{34975}{32097} & \frac{181282}{32097} & \frac{48033}{10699} & -\frac{145679}{32097} & \frac{7747}{32097} \end{pmatrix}$$

@ Exercise 8. Consider the matrix

$$A = \begin{pmatrix} \frac{4}{5} & \frac{3}{2} & -\frac{12}{5} & -12 \\ 2 & 3 & 2 & 4 \\ \frac{23}{5} & -\frac{1}{2} & \frac{19}{5} & 14 \\ -\frac{7}{5} & -\frac{1}{2} & -\frac{1}{5} & 0 \end{pmatrix}$$

(a) Calculate A^{10000} directly.
(b) Verify that the eigenvalues of A are 1, -2, 3, 4.
(c) Write A in the form, $A = P^{-1}\Delta P$ with Δ diagonal.
(d) Use this formula to recalculate A^{10000} and compare how long this calculation took with the time it took to make the direct calculation in (a).

@ Exercise 9. Let $F(x) = x^5 - 5x^3 + 3x - 7$. Notice that $F(x)$ has degree 5 and that the list of the coefficients of the terms of degree *less than* 5, starting from degree 0, is $[-7, 3, 0, -5, 0]$. Consider the list of the opposites, i.e. the list $[7, -3, 0, 5, 0]$ and use these to construct the following matrix

$$A = \begin{pmatrix} 0 & 0 & 0 & 0 & 7 \\ 1 & 0 & 0 & 0 & -3 \\ 0 & 1 & 0 & 0 & 0 \\ 0 & 0 & 1 & 0 & 5 \\ 0 & 0 & 0 & 1 & 0 \end{pmatrix}$$

(a) Verify that the characteristic polynomial of A is $F(x)$.
(b) Generalize the construction of A as $F(x)$ varies and verify the same property for the following polynomials

(1) $F(x) = x^{15} - 1$
(2) $F(x) = x^{12} - x^{11} - x^{10} + 2x^7$
(3) $F(x) = x^3 - \frac{1}{2}x^2 + \frac{3}{7}x + \frac{1}{12}$

Exercise 10. Let I be the identity matrix of type 3.

(a) Prove that if A is similar to I, then $A = I$.
(b) Is the same statement true if we substitute any positive natural number for 3?

Exercise 11. Consider the family of matrices $A_t \in \text{Mat}_2(\mathbb{R})$, where $t \in \mathbb{R}$ and

$$A_t = \begin{pmatrix} 1 & t \\ 2 & 1 \end{pmatrix}$$

Determine the values of $t \in \mathbb{R}$ for which A_t is not diagonalizable.

Exercise 12. Let $A\ B\ P$ be square matrices of the same type and suppose that P is invertible and that A and B are diagonalizable using P. Discuss the following theoretical questions.

(a) Is it true that $A + B$ is diagonalizable?
(b) Is it true that AB is diagonalizable?

Exercise 13. Consider $A \in \text{Mat}_n(\mathbb{R})$. Discuss the following theoretical questions.

(a) Is it true that if λ is an eigenvalue of A, then λ^2 is an eigenvalue of A^2?
(b) Let $\lambda \in \mathbb{R}$. If the sum of the elements in every row of A is λ is it true that λ is an eigenvalue of A?

@ **Exercise 14.** Consider the sequence of whole numbers $f(n)$, where we suppose that we know the initial values $f(0)\ f(1)\ f(2)$ and also that we know that the numbers satisfy the recurrence relation

$$f(n) = 2f(n - 1) + 5f(n - 2) - 6f(n - 3)$$

(a) Calculate $f(3)\ f(4) \qquad f(10)$ in terms of $f(0)\ f(1)\ f(2)$.
(b) Imitating the example of Fibonacci's rabbits, construct the matrix A associated to the given recurrence relation and calculate the eigenvalues of that matrix.
(c) Calculate a decomposition $A = P^{-1}\Delta P$ with Δ diagonal.
(d) Using that decomposition calculate $f(10000)$.
(e) Decide which values of $f(0)\ f(1)\ f(2)$, will make the sequence $f(n)$ constant.

* *

This finishes Part II and with it the mathematical content of the book. The strength of linear algebra has only been partially revealed in this book and I hope that our readers, arriving at this point, will not feel totally content with what they have seen and, for example, will try to understand the material of Part III.

Part III

Appendix

> *to make difficult things become easy,*
> *is not easy*
> (Caterina Ottonello, 9 November 2004)

Problems with the computer

As was already said in the introduction, at the end of many sections we can find exercises marked with the symbol @. In order to solve those problems we suggested the use of a specific program which is geared to such calculations, namely the program CoCoA (see [Co]). The reader should not expect an exhaustive description of this program. To be totally honest, I don't want to be exhaustive in this description. In fact, the goal of this appendix is only that of inducing the reader to consult the web page

$$\text{http://cocoa.dima.unige.it,}$$

download the program CoCoA (which is, by the way, free - something much appreciated by the folks of my region) and, with the help of a friend or of the manual, learn to use it.

A simple way to start becoming *familiar with* CoCoA is to read the story [R06] and the expository article [R01]. Naturally there are many other programs which can be used, but in Genoa such a choice would be considered... traitorous. In order to avoid this danger we'll see, in a bit, how to use CoCoA to deal with a few specific examples.

Before beginning I would like to make a few remarks of a general nature. Everything that you see written in **special characters** is precisely CoCoA *code*, which means that it may be used also as input. The parts of the text that begin with *double dashes* are simple comments that are ignored by the program. Now let's start with the first example.

Robbiano L.: Linear Algebra for everyone
© Springer-Verlag Italia 2011

Example 1. Solve, using CoCoA , the linear system

$$\begin{cases} 3x & -2y & +z = & 8 \\ 3x & -y & +\frac{7}{2}z = & 57 \\ -4x & +10y & -\frac{4}{3}z = & -71 \end{cases}$$

We could solve it by hand, but let's try to see how to solve this system using a computer, in particular we will try to solve it with CoCoA that, in this case, will only be used as a *symbolic calculator*. The first thing we have to do is write the system in a *language* which CoCoA understands.

```
Set Indentation; -- write one polynomial per line

System :=
[
3x -  2y +   z - 8,
3x -   y + 7/2z - 57,
-4x + 10y - 4/3z + 71
];
```

To control the content of the variable `System` it is enough to write

```
System;
```

and you will get as output

```
     [
3x - 2y + z - 8,
3x - y + 7/2z - 57,
-4x + 10y - 4/3z + 71]
-------------------------------
```

Moreover, in order not to always have to write `System`, which as a name is expressive but a bit long, and in order to keep the input to the variable `System` invariant, let's give it another, shorter, name.

```
S := System;
```

At this point CoCoA knows that both `S` and `System` are the names of the system with which we began. We can modify `S` as we like but `System` will continue to be the name of the system given at the beginning.

Inasmuch as `S` is a list, the expressions `S[1]`, `S[2]`, `S[3]` represent (respectively) the first, second and the third element of the list, corresponding thus to the first, second and third equation. Now we will use some of the *rules of the game* i.e. the following elementary operations.

(1) multiply a row by a non-zero constant: `S[N] := (C)*S[N];`
(2) add to a row a multiple of another: `S[N] := S[N] + (C)*S[M];`

```
S[2] := S[2] + (-1)*S[1];
S;   -- Now let's see the output

[
  3x - 2y + z - 8,
  y + 5/2z - 49,
  -4x + 10y - 4/3z + 71]
-------------------------------


S[1] := S[1] + 2*S[2];
S;   -- Let's see the output

[
  3x + 6z - 106,
  y + 5/2z - 49,
  -4x + 10y - 4/3z + 71]
-------------------------------


S[1] := (1/3)*S[1];
S;   -- Let's see the output

[
  x + 2z - 106/3,
  y + 5/2z - 49,
  -4x + 10y - 4/3z + 71]
-------------------------------


S[3] := S[3] + 4*S[1];
S;   -- Let's see the output

[
  x + 2z - 106/3,
  y + 5/2z - 49,
  10y + 20/3z - 211/3]
-------------------------------


S[3] := S[3] - 10*S[2];
S;   -- Let's see the output

[
  x + 2z - 106/3,
  y + 5/2z - 49,
  -55/3z + 1259/3]
-------------------------------
```

```
S[3] := (-3/55)*S[3];
S;   -- Let's see the output

[
   x + 2z - 106/3,
   y + 5/2z - 49,
   z - 1259/55]
--------------------------------


S[1] := S[1] - 2*S[3];
S;   -- Let's see the output

[
   x + 1724/165,
   y + 5/2z - 49,
   z - 1259/55]
--------------------------------
S[2] := S[2] - 5/2*S[3];
```

The last output is the following

```
S;
[
   x + 1724/165,
   y + 181/22,
   z - 1259/55]
--------------------------------
```

We have obtained an equivalent system *which is much easier to solve.* The solution is thus $(-\frac{1724}{165}, -\frac{181}{22}, \frac{1259}{55})$. We can check it with CoCoA using the function Eval that, as is indicated by its name, evaluates expressions. Recall that while S is changed during a calculation, System always stays what it was at the beginning.

```
Eval(System,[-1724/165,-181/22,1259/55]);
-- The output is the following

[
   0,
   0,
   0]
--------------------------------
```

Now we are really convinced!

The next example shows CoCoA at work in finding a Cholesky decomposition (see Section 5.4). In this example CoCoA will not be used only as a symbolic calculator but will work at a higher level than in the preceding example.

Example 2. Consider the following symmetric matrix

$$A = \begin{pmatrix} 1 & 3 & 1 \\ 3 & 11 & 1 \\ 1 & 1 & 6 \end{pmatrix}$$

and let's calculate its principal minors.

```
A := Mat([ [1,3,1],
           [3,11,1],
           [1,1,6] ]);

Det(Submat(A,[1],[1]));
Det(Submat(A,[1,2],[1,2]));
Det(A);  -- The outputs are

1
---------------------------------
2
---------------------------------
6
---------------------------------
```

These minors are all positive, thus by Sylvester's criterion (see Section 5.3) the matrix is positive definite and, as a consequence, has a Cholesky decomposition. We use elementary matrices to reduce all the elements of the first row and column to zero, except for a_{11}. The reader should notice the following neat way to define elementary matrices.

```
E1 := Identity(3);    E1[2,1] := -3;   E1;
Mat([
  [1, 0, 0],
  [-3, 1, 0],
  [0, 0, 1]
])
---------------------------------
E2 := Identity(3);    E2[3,1] := -1;   E2;
Mat([
  [1, 0, 0],
  [0, 1, 0],
  [-1, 0, 1]
])
---------------------------------
A1 := (E2*E1) * A * Transposed(E2*E1);
```

We get the following matrix A1

```
Mat([
   [1, 0, 0],
   [0, 2, -2],
   [0, -2, 5]
])
```

We use elementary matrices to reduce the elements of the second row and column (except for a_{22}) to zero.

```
E3 := Identity(3);   E3[3,2] := 1;   E3;
D := E3 * A1 * Transposed(E3);
```

We ask CoCoA to tell us what D is and to check

```
D;
Mat([
   [1, 0, 0],
   [0, 2, 0],
   [0, 0, 3]
])
```

```
D = E3*E2*E1*A*Transposed(E1)*Transposed(E2)*Transposed(E3);
TRUE;
```

Let's put

```
P := E3 *E2 *E1;   TP := Transposed(P);
InvP := Inverse(P);   InvTP := Inverse(TP);
```

and check that

```
A= InvP*D*InvTP;
TRUE
```

Now we have to introduce the square roots of 2 and 3. How can we do that? In CoCoA we cannot directly write $\sqrt{2}$ or $\sqrt{3}$, since it is not part of the language of CoCoA and CoCoA wouldn't understand those things. For now we have to be content with the two symbols a, b. We'll see how.

```
Use Q[a,b];
B := Mat([ [1,0,0],
           [0,a,0],
           [0,0,b] ]);
U := B*InvTP;
```

We ask CoCoA who U and U^{tr} are.

```
U;
Mat([
  [1, 3, 1],
  [0, a, -a],
  [0, 0, b]
])
----------------------------------


TrU := Transposed(U);    TrU;
Mat([
  [1, 0, 0],
  [3, a, 0],
  [1, -a, b]
])
----------------------------------
```

The conclusion is that $A = U^{\text{tr}} U$ is the Cholesky decomposition where

$$U = \begin{pmatrix} 1 & 3 & 1 \\ 0 & \sqrt{2} & -\sqrt{2} \\ 0 & 0 & \sqrt{3} \end{pmatrix}$$

It looks like we have cheated a little. But, in fact, *we haven't done anything* because we never used the fact that a, b represent $\sqrt{2}$, $\sqrt{3}$ and the real reason for that is that we haven't had to, up to now, multiply the symbols a, b.

But, if now we would like to check the result, we will have to teach CoCoA to make the simplifications $a^2 = 2$ and $b^2 = 3$. We can teach CoCoA that by using the following function, which CoCoA will understand because we will write it in the CoCoA language. Let's not ask too many questions about what this means but rather let's be content with the fact that it works (like when we buy a television or a cell phone and we don't ask ourselves how they work but just learn how to use them). However, if you are not happy with that answer then please head right away for the web page http://cocoa.dima.unige.it and you'll find all the explanations you want.

```
L := [a^2-2, b^2-3];

Define NR_Mat(M,L)
  Return Mat([ [NR(Poly(X), L) | X In Riga] | Riga In M ]);
EndDefine;

A=NR_Mat(TrU*U, L);
TRUE
----------------------------------
```

In the end we get the answer **TRUE** and we can relax. It's true that $A = U^{\text{tr}} U$.

The next example shows CoCoA at work in raising a matrix to a power. This was the problem considered in the introduction to Chapter 8 and in Section 8.3. This examples gives a practical way to deal with that problem.

Example 3. We want to raise the matrix A to the power 100000, where A is

$$A = \begin{pmatrix} 2 & 0 & -3 \\ 1 & 1 & -5 \\ 0 & 0 & -1 \end{pmatrix}$$

And now let's see CoCoA at work!

```
A := Mat([[2, 0, -3], [1, 1, -5], [0, 0, -1] ]);
I := Identity(3);  Det := Det(x*I-A);
Det; Factor(Det);
```

This is the first step

```
x^3 - 2x^2 - x + 2
------------------------------------
[[x + 1, 1], [x - 1, 1], [x - 2, 1]]
------------------------------------
```

Thus -1, 1, 2 are the eigenvalues of A. Now let's calculate the eigenspaces.

```
Use Q[x,y,z];
L1 := LinKer(-1*I-A);
L2 := LinKer(1*I-A);
L3 := LinKer(2*I-A);
L1;L2;L3;
  [[1, 2, 1]]
------------------------------------

  [[0, 1, 0]]
------------------------------------

  [[1, 1, 0]]
------------------------------------
```

Thus, a basis formed by eigenvectors is $F = (v_1, v_2, v_3)$, where the three vectors are $v_1 = (1, 2, 1)$, $v_2 = (0, 1, 0)$, $v_3 = (1, 1, 0)$. Now let's write the associated matrix M_F^E, that we will call IP, and its inverse M_E^F, which we will call P.

```
IP := Transposed(BlockMatrix([[L1],[L2], [L3]]));  IP;
Mat([
  [1, 0, 1],
  [2, 1, 1],
  [1, 0, 0]
])
------------------------------------
```

```
P := Inverse(IP);    P;
P;
Mat([
  [0, 0, 1],
  [-1, 1, -1],
  [1, 0, -1]
])
```

Let's check that A is diagonal
  by using the matrix P

```
D := DiagonalMat([-1,1,2]);
A = IP*D*P;
--    TRUE
```

The following function (which we will define in the CoCoA language) tells CoCoA how to find the power of a diagonal matrix by simply asking it to find the powers of all the entries on the main diagonal.

```
Define PowerDiag(M, Exp)
    Return DiagonalMat([ M[I,I]^Exp | I In 1..Len(M) ]);
EndDefine;
```

Finally we will verify, experimentally, what was said in Section 8.3.

```
R := 100000;
Time U := IP*PowerDiag(D,R)*P;
--      Cpu time = 0.02  -- secondi
```

```
Time PowA := A^R;
--      Cpu time = 10.29 -- secondi
```

```
U=PowA;
TRUE
```

The difference in the amount of time used is... monstrous. However, congratulations are due to CoCoA who wasn't the least bit frightened by having to execute 100000 multiplications of type 3 matrices and even did it in a few seconds. If, moreover, you looked carefully at some of the entries of U, for example the one in position $(1,1)$, you'll see that they are *enormous* whole numbers. Would you really like to see u_{11}? If you truly would, look ahead. This is brought to you by CoCoA.

```
9990020930143845079440327643300335909804291390541816917715292738631458324642573483274873313324496504031643944455558
5493001879966076561765629084713542474928751988896298736710932463504273731124792658002785312410887370856052872283901
6456869102685067592351791469705285764469680152483234547554325029278652080696577709717411022320429763512053307799689
7925116619870771785775955521720081320295204617949229259295623929657978735581689765254957973131448062492602618379 41
3050805826860315351341787396228349908863577580621046066363721305877953223449720100808486369541401835851359858035 6035
7402187290815556658060718646126897283979462184226757934693889335724758876195913765762411125020708704870465 1793 9639
8710109200363934745618090601613377898560296863598558024761448933047052222860131377095958357319485898496404 5723 83875
1707022423326334368944232973818777331532869442179361253019078689036036632831651502726139934152804071171191 4923 9033418
7493539445589630129219725641771723354354475155237931089226818240245275575209470464218594386286563274423133 2084 74222
1551493315002717500642288262118225649349600557457334964678483269180951895955769174509673224417740432840455 8821 09137
9053756467721399766217852650571698548345624875183223832503186455057211143699341678981570617025512281297806 5194 8692954
0533915465747994129749919034850754433641450563165739600669338242731643403958012128026098421224751420783471 2224 83141
0304068603719640161855741656439472253464945249700314509890093162268952744428705476426472253167514542118223 1455 388374
3082326422006330251375331293651643417252062561553117947386191429047614456549271284181751835313270529754953 7056 14382
3957322793967303010607745684847742783219534922798383643616376474296954590667236912413632590321233564313589 4465 21910
1882123829740907916386023235450959388766736403229577993901152154448003637215069115591111996001530589107729 4210 32230
4242620356934932160529275696258584458223545946452769231081973058062803265167364449343761732409753342333288 9730 2829591
7356927301328642331175960523049517167703316370952225695246040214338765519764401652814802234833188109755942 1960 47647
9388520198541017348985948511005469246617234143135309938405923268953586538886974427008607028635502085562029 5494 352480
0507965215649196832651067441009678229519541616177175429975200098873073778762106858907709694116104380286239 5044 53247
8959187076028926039348982610077488767285291810646848914389364906478459121161219330070790053705904218801285 6559 40369
9070888032966871611655961232331998310923228662180321880439447572986762096935819784385927969250123326935194 6932 07
7243355273655662482237878338880749992768316334403186044636187037897843130328438234704109443065914719283411 9097 51852
3921232767438499056156368843293903944200261753097685060513293710144908639614162055605354733556992670094137 5271 82914
2407234267937565069765564741593410131022534283008040907958733294542135513073020501715984242307604692097329 0729 01 41606
3539608805592023573768856478522400927771114891344924169956071717862984365339781808694741067511113535237115 4043 65993
1088969748565880088786197493435792924620405176724601125061840401196628987267380307049836121797448467910074 7846 35619
4664829224736134115135567179291781968056053726484141128347858241259121954601184412409349782963317042002530 4186 61694
9623187358606524854102222186954422378828918971208051457514136196480536972316457056499847953765717454812859 7406 0773
3915877533235521560943591927519935101422224696301701371741933750491925363295101115292951836282819191821651 6764 5594
6515828048984256116748150367805267878662716999649296949377045794876146628110929982020737013330324451005385 3785 51188
8034741481986651145793226849009930040937361685552941734420599253719652449979254831593437063439703718096114 7032 30741
8698503505472228902717485033336832830028113291084169315045738993318393459329294994279601530975611870891892 9528 44907
4243284767006243171171622731766606796101967802204564589015899524704741001158110963633731329388356868949408 7593 34176
9093878063985846473005889281759988444774861300631530687600700848372675277897773568300427789027721056838330 2147 02797
2859533633211056406426390972457994968616290801960414175393576887658799242854992364388686814247047841 7843 5
9541893241450987505759403013249697541696955330296880219304874163501097920036210238768275176369980977614979 6360 96704
3481401241306835768799049974365962964957054595247353820003637703248949821033313329135623151698544104153170 5419 39282
3472339884845355217320368808831210094394143299456502815307510870986046812248029738364254286252441449893319 6529 620237
2608586509050307993308652001231671915182765742095689513136184095412121473786311042897717861448158316965848 7669 49554
8262525049612270447147122296202746823629098037746937698735894212544179235529838747963409525390978873346973 2603 0975
4415647480547373273276724865275903499533635412695390045885498868357492786461525204080049011478589228908544 3353 99699
4780867471613519785838571456421583171193004117989440790268346357650339888087265121788357729762649921382743 6573 9927
3022387925769242327854872012972553860719683037824830637258998084846385038283562584039173118726943814465536 5169 00625
3002321759134308475521590147529914921529694443623669108332336937679931382092758700242462383312182367152367 7209 841718
7703860172308522440431763336027597331612012524832308532928898615455922142737850741097882244729512663572255 6716 9
7794097673415430172892683326350774512101678691213344656807397973727114619192999381181788275414217929268837 9028 54309
0994244126051194584923790996632955026386570111488414226616296981007346521092850457940861508094054577978643 0150 6899
9958634164700528220562786008864025709432444254044034243140203812074857537999016066465520986980790589347320 2430 50635
9073638215212806000418275293254852437259857420955463280338303324282507115188017756337398115237619946862632 7055063509985125433387559460154090086201429362567373833169308232885432700148774663511883308851737752688195 2636 01653
```

```
4490329965893251199650410965367461891486159948163204089193057723863039631185821334133711009638911383659689591471537
0925073998461682046426447290788976525593505136546978364603138820619560578517561504972661817649030304982138543738696
2122346261140430356009670425470123173604497246232874525751511987718015857428293890256508259882754951108654247042183
3726402307804568165142051780741819609640151346176079436276961222812611861091276681488050095096388903287771083765105
1900076128058473969258768737937306664751387942217354694021157675557568970168734104342446525522568974329716152742558
110503495045718931752447070410307760830365536871418038872360294887280555907527111155907947569269039785196019397903117
6807035680194493610685064056851929064504868553562825678722573454414655654118781671772985061287404462089071850210851
8025052924590359814117522720320552642597751984410742492179242039080014606225999422109717176118746845802673724801365
60386690997107134725585972321702755405508508209041898753482922200417899847503051953717906200150933302330238818065191
8240555081867216471170230752992265222803382040411338662533581504293411514398093998641636563392362067387425934271344
4701242702722227197573203194489407863655511639619115985907995399083680129468810771595938084908111251938016414866250
1410952866809148285031239389609976591759773154327971739457625603650235879315599261708523150742478498142565646930081
0506197639739559173354547291752673959879011747744921774577190816948954379031457815266738968941060458835144502613064
5637237687631129964576699475767340673583537218704935177320214779403972566532581731659202199752942824432778102107532
1605801044321210672082738761007783324265696247656210631269754915462243439780612539993138939157820085600111713197734
431304129982156269850988957227781596549784598558722345219917784196641066221220490390178867379979062572055230
2412002780864726282517525092332555378737792434915926182736151242592242725872699844014132567954640475742451126102857
3941934797163831871383707227824226193840210108962712816857322877642102987088955571483977434974181098496336339103977
7825422517940002214348588620753212264661361448751873514249446957583674478502801319303390194973871616311380086409340
852929772974146283614220112057302742730956665884988499651342951882879370161474995046851851651709758146686994243673
14003699238123248392061862866303740235133390777190745522424851574872613607504802096097656786202523235627302055570543
8672918925557167239687199166965183473698744029594239522086348134148402983893948559332725662731819413773545189154438
4960966451285561476772564932516923822003229733483331726306200059192067441136913243720403578098676440894672367334459
9039052836824114220118869492653245313993237641005639967442736936065868512439367371553496963589704707062467430646815
115255807723671293586915466771179280674603197443660223566481191117532423979227160321393269090043143357251193559178578
7438931910836722219749594385217682817205537433987939372420698422586002026702204502341484432231418159837133223467338
18599672021521229990668553185937651881228156443498428978040444742531731871982498802190561247568350531219366840911136
1345174038040004128147452745368753012274122629254943100485886808262716931289141423746697321568830696259629782428682
8479758198382340345222708757345393897126910174835179971320922911746435455023926629385455742630840473245989374307836
8758640879143123192283073049753624817085798498705503143573686604535040996796853898025784779104192310525051446968852
4872634436536945392137219471599815662445433056279828277734958466675701896990098511016449104799360292296164513299522
427140490818654980808181447654414760708349431799213476932057754937275811469799477819807044909830887802418354528095576
2681782215964107122165099951969619323687669599243369390186758304779513485649674966430069785481208610130290264826404
02316621018258604172734501970838570222765185859039312362724523200653598390760412789800762822079124616322271305190667
8029644273228763460060289213316094057396003523608149487855784819923604033709243920895785100953666245538047118419834
068589948934026518132741491251925687298248779829383517102243898868347555622182148660755186284205084412408550884
084994052265485357766788644448605043012859884699155016926211730495502683654983406379433415847973906669732718175611
11234452127011382877331707036488435606925350128577456371660129278461941357564915082693230100636559087991831480179606
66489360686487556915470086060270390964009051460936051303728098498997647293420597107610436570667036963673130686617730
6436133318416433210407347921077568870375419324110176448807167330417626803452812794672924445291916188209533594566715
84509414763015370325881092549216206024239983839396395708642683155577379235428304771690653850721939927266830567227442
9137526802371781759084372697807080996901926950025924210204851939538051551586632664182304521293710468402188066651652
3143859749808162142014805135513186545301487129824993862427215434537239180221512611585486197539898311088819635887556
57935933105978959920532404396845908621932320152322576689695094153983937213087472477329444930535757577694362680332816
0550603532196850012071951916714260633640561790062186716046386842490672475728557640716313978246391886937896352477082
32195940040045082959853651517642610129335911478337995658501766175786429968683741724783883864589231992004119639092535
4226074199412181950196251598137076467085022479201649174056794994024967593003129957099742026595785390010668925864718
83902605841758236944294710952294472987418327395229247729093779983495276999514512881843121031747660005057328130456
9814187673870583558968056881206290167238970760303954970827341848406903721517986226658192955542075569565413997681423
50274905946625927709860445887936755621437096487057446653198147728921177290937799834952769995145061157204394128687121
538756842568036622321369508041915737469009917048039885947260486258684449767623623187240853395002049892959636187394374
0891888856890369112898783232589260133115526090113191196253652938464399346566649027110852283447018476839157362389523
97327811583993551780220761774507559291383814685265773091522905029145241134006895661073761320484545645543610238773
08759430840869978271131542551347205893939450135481352441767541313850127700241562619361696497099441569335416665909
3575978477759464689539235162297236743654360805861862098881267709889294476965436032466887442974013036500977029160353
6426502844618046128137324006034947889707723262364927515545013731797277615272236008596598736693841467685030166103516
988390995570659331132195740032510468362952763169728939909450912796453617925934868184558301686342806532050287
2910502690664372617066177843931115473714975909611614449381210674603725145261627161521293504314112415584984757409187
8458410177325030553207237661933970254488335435069113577757765584551136714634629223114497809546405079026755854160954
2605577834064673561893985031603030466875225797083879932987691829868229706673363246647939654535442683864043226502
960288812985572421981872080402072960774961288816677328883862060997792541682215671370366739133332208652087190447304
859301746679357817383083046945850718772674476903765231970142636502172093102619263046542112708094769682392207383
7974820570347931965251392213701998775439311786012381989247070476295875015162999128929330690253435754882514549484693
283228982029565060690330847048251774413116646477457651805534346885804098713227345308183679456436443015254526427
415810073106932308073490159051700875365416960099891156959845470292662937404939367768528572303975469118021469267219
333846563053673846537824970049506209403506060624585841262790238424278528055159925318826278627153198956716539838543289
09311577122438618119724985644233907558812574859217681794972134183331728971497769498211722733766987836121278846528934
9470676995146785835955074121568375032937822352378665103266923449887690619794116574936418347914823983988002067649967
28921498384662446873774228643856306442522346121063996241816688218463748633375227290779942089605452648638625164117180409112
903451378140021506899966762080732651380427792460930762263446990941591932231425650299964031343835640087766689537759874
7340312047709188659635557994464368820598124204047780294380393245972220754173958959669203405216168083758892163604858
478992966390310734994566702783061048313086287268066101530074552166989385733872690975026078691935349057026830281281070
4750603181709274434806504843422134152205157962677639364269816885148554520688978020365737496043584180735523092410264794
4364239514748585398025936701720558876798530997104329040488446754170191891227574857524021217448903266811606478397850
0463986419364579208210526605703921195311457592621400830660301768120178560328650220171141381164215598670263753115555
482195838890539354459218058727243374989924204154281666821846374863337522729077994208960545264863862516411718049112
903451378140021506899969697620807326513804277924609307622634469909415919322314256502999640313438356400877666895
7340312047709188659635557994464368820598124204047780294380393245972220754173958959669203405216168083758892163604858
478992966390310734994566702783061048313086287268066101530074552166989385733872690975026078691935349057026830281281
0638783700836085633247740114266045887977128433350893417521208431335491558059452175037667281368189897010757950349759
311471138511775290847083826469309436638378518489798723688771534770333456570254004964039988776074595127614593669469550
8877796318400093229973736770518512573233230127134210937086868095026430220585108728711416356017162762670441061899681287
2133264222683778553278878223102045397622725830149026097109311949956859508435945092588033630195518918462129909149201
5583829860756819073737595857334826171379470359250397212631431619215957146974043134602128424464336601284246453826539
244231676741103641755777699188671991051145690200290577169917749779483594989810115650557253501242519595931483440417116
2222055045896716367544786348838550027521149752624644112201077486233347624641825765971784039804366176695605670499
3678935037089317461756028207467905811080628034889313676877313323472117854103828921308058036930285391038470452801834
4058313252744129928355732704028220901951560054449764635932065083760864473989692709381120504251266930834839999956860
```

2163561489401674425640784693501527302457494357222050094036808103211368989464430620416598783604068113312171335049605
0494416547288392637860786838573151536637341631256612260323597774247394032479347874077035272543653245530843825086569
3864330020957120719122826689237609273524467576647981797622831709248187143189474449928329051087066866263901462754600
1866835836220624455538288671436214157736382448525706646534230020032003386087185683387095282564231892337305827467825 1029
0523921446411759070008286287491564721296222310014984277391058526271209570263342513906927554124992582689875591304848
9059083665047386569005980726512682643809800080768716889749767005747228285435364293426469423701296901225200041775975278
0993413991494556756554157912890141418071809708180106686260003347467834921163543307417665561978654129443315796108567
3390721097081524769436020823107737731925593559191936150999224185186976251182191759844939642292332225869395832058493 5
7600173388242916813257003198490942805178528782842560021549233250032941372842601110962343206148554757903880284922813 9
3971309791805624477034942995305859094150312374002134454316070495388091178705710456862116210696550230288866170603145
8022288434490687940489074406716455756516059639748166074457540940089191108394512623399045336477254174113312181611537 5
8983077390402549739751620825837898384996200874749622331436354553599710472896396262914675542281735624665804212007275
4012739822696518025961303343297928653300636541756760033696882439263912807131230387600760399572856583333390838952342
4933715307045747148274355103741714702400686741914187439823044920277200968228247464080277906810474160288261344080325
6607530230172031981070796893051954145239327661783905828943932146148894821695918579498502263858318704473290299790473
1937538035439202857063506724842450089206897129953082137473352557929825658034697795049740533295062205845869767823765
4735809267075389173665240157752589020235488322748067896132846682186597286978559267282682265645757795867057109772839
6605101516387163615321918203652729897586807786443044896628670393188486921931839182927172554503767733076856308657600
6653426222545350576290308627189373643274480835589282201023749515346652628575042229509954460144288018590614433796791
4271741327828697599825999381908447746977522056789839537857015460917695006128689610825519373486612568225010585752 37
5050450503641627804406676905130564807191014132285063482665726828511933059433717373896236547155348165916300868385 20
9054910601945967582237019115135811420715212206735219487770695216085317174775805425541380389788737160582417719042524
0105637066179299639580870671708387295023805723854761049506173268958794824031956949900059459371984900410455003441501
4066154526994609094441391439071866309009249691812468813088126242358787204856089608234037677437509496768451058059579
9771890066053549035996178424512109614487706318746691029235662771116520352264701616629648008380899788768776930822 08
5653006815394085674804134946949187070499266021006645581983288613067802146036989931744999391072312304050883424853 8379
6742208847060385334943928147519685791292174441917386398739155069181398492193183918292717677538628240600245182114 90351601
2449289394057849405543217358977950386779244861476944873467186358381729662707334140402763102982156230314215774337929
4616649374361580459093606029858122457608525327005700618776095827459794967405315882162695982851143006320897730473133461
17214294580540717887622301146255851052093076177149327860354126271988052282686410705504936437055581694583084065096 04
4732186826685821690378486803228047986402170631565368748929046767901586470937947876237301382202637378882282756416 372
5872670921602434077279298565679825234946428376162918717556993454693724579744644374586123577768842647346924365909558
5531625715648079690340526715047607547426769600382558140343471652083459864538432593908154392918939089896109083 5219
9251905353039835908151682924757104335478642495498074138874538630250100574401503129413082450564954394374386907595891 21288758000
7804082911975207095278133335550232776761783391381794105109412999680430103899475051243232331431120663598726618610084
7172546969087266374568584666461620932234276806846017185167024276711259130800593583653816003599124447685286534985 3
9701387608081702091847286335136878374525122923629243731377977941681631309836888336245924882700425332753101740374018 5
8155192663147588032450316349828944090941459435709839331454077937638533884824056964583661906580362312059760037127 664
8309951258591021429023730605200088345157497244050288634025006877054052336885410685938903966237991377648571577 20937
4968105571827778142965924425166670311013582062531651890693646266361337678677001574909561444962828205690820330013 12
1289873007855206280915035880334071746233511114882911788112711788259641473422758858031624736482535088797676299 074
8132426924617141695210685488182753616127869552164123832632461640346027710530522171274169098317144258427543593251 937
0836792052487018621270857453151819707979794216322369627530303625010057401503185241771381716916959066550480494267 47
3696656051819527729744044078248741635308707480344712630755045301876873451163126000257176570829411344715905 83268 59464
0229754890618685697155293253336791929045091813389890395312261140776041463413951202531990687206904506543173607 993524
3479108203969547806192646532546440531246586362020616959788832641991034651974219439600618174594673322113671 1305167
5844970722080638249309285348257747082754939860608508486386549912912220391895379332321814344998458680557267915553525
2121139077233890379794274852093095046415996391960532960302927392746639063407543902850032647016166727391709148790514
8812821582894579949748623433452865460274945133477267345362799684706610937707688048379567860499887327287273158785195
5052169469865717593728759845149940873667356996903222197283739213789428737148328927877210957124703777285270451
8165355628562165676197938296083725208475920318111438440400511461399032627336214615066785188007932756348644480372696
3278475818453446850956653653807238077367933526242145533028833679270741928388726854849773873066700075519657614315 83941
1020424926231332427702708217100471371521679487246874118929805623910487138975034203376673344989060052111728446705
3052132742533218973497888830794382064452609912556974130835943046150712876440962232039557455150019442996232739317 09148790514
8812821582894579949748623433452865460274945133477267345362799684706610937707688048379567860499887327287273158785195
5052169469865717593728759845149940873667356996903222197283739213789428737148328927877210957124703777285270451
8165355628562165676197938296083725208475920318111438440400511461399032627336214615066785188007932756348644480372696
3278475818453446850956653653807238077367933526242145533028833679270741928388726854849773873066700075519657614315 83941
6721403439803621918243177242110484823901622114332463945118321907370362763625569475353726194361269966453422295 8906
7369520148703938865015087074414735220220770411070088328030731545041085721762448600324574529642445801550887535 5814451
2891817368634611656711105015110971476054436153792020754611507352916793535980374623315803540250729358542213495 023965
5004818456256327794930441926282941363637022415075813444776889555669616251741423859274176434749486631831 82648
2100837857490802221991221446941174002708561590951048858456793923326023018437403287028085697823111824603926421 290247
2919253522678756789786856460057054510894531419064937156889687873486982693386634574807801078719254129784 90715527
6683405362407228975018016417486695226639468897799829747034414846605232977531914266384739068105416711680 06131262918
8347648006396547623451925405804002577279508590618173260001883566775762496350845122664105575516451344702 620384548 07
7617370192362640638483137184208747722309446797866122164968984232278924876260516789686457539012260361 0679246617 2839
0016267344059903224026536343550301552983264519865881303942586617734987280231892544278016113318354 27610325800 2393942
6078587569494851193007158539298431563967575531352484692729730715545710942342345178597052363567 2865515162
6567238924808065927944895307870815360666741314598260798811287845631300778574348418961080 9499194360375610753 2635
9414143476882494692530787790693275290895909945065272071979419682741840819830567497688086236 4608193456236377 02467815
0731334661979231199391171959793528331174968205005772999385664769842608151894086677632085548 31598361888504 483399113
5150513290970399008891906313337580409488695589546591246084815046213202957333759928989593 29664444286890068 47453166187
8821297194146191914132978673375283702995190706179924178140249356285006536273771636355569 72000324699613066 5282451695
1672731618313938675587912782649337580400110592113722708391235184597091074806698526960809 0918187298501574 0645445482
6447107863338655991131888101377463189844967459890115403342593784700821107348184715212533 745072715381310 48296 61297
9081477663269791394570357390537422769779653811568292232084025970251553047343898831 09376;

What have we learned from this example? Surely we have learned some important things. One of the things we have seen is that numbers can be very, very large (the number above has 30103 digits, if you don't believe me... count them!). However, modern calculators have little difficulty dealing with such numbers. We have also learned that you can get to the same result using different roads. Sometimes one is simple but slow, sometimes another road is much more complicated because it is based on non-trivial theoretical considerations, but it is fast.

You will have to excuse me for the fact that I have contributed to the deforestation of the planet by writing that enormous number above, but I hope that the motivation and the conclusions that we draw from it are worth the sacrifice. And what about the reader who would really like to see u_{12}? How many pages of output would we need for that? If that person were to try and make the calculation with CoCoA they would have an interesting surprise... but, perhaps if one thought a bit...

Now that we have gotten past the nuttiness of the preceding example, let's conclude this appendix with a more complex example which relates to Exercise 9 in Chapter 8. Here you will get to see some truly noteworthy characteristics of CoCoA.

Example 4. In Exercise 9 of Chapter 8 we were asked to verify that every polynomial in one variable is the characteristic polynomial of some matrix. Let's see how CoCoA can help us with this problem. Let F be a polynomial in one variable and the matrix we are looking for will be called Companion(F) and we are going to create a CoCoA *function* which will give us the matrix if we provide the polynomial F. This is how it is done.

```
Define Companion(F);
  D := Deg(F);
  Cf := Coefficients(-F,x);
  T := Mat([Reversed(Tail(Cf))]);
  M := MatConcatHor(Identity(D), Transposed(T));
  Return Submat(M, 1..D, 2..(D+1));
EndDefine;
```

Now we define the characteristic matrix of the matrix Companion(F) and we will call it CharMat(F).

```
Define CharMat(F);
  D := Deg(F);
  Id := Identity(D);
  M :=  x*Id - Companion(F);
  Return(M);
EndDefine;
```

Let's do an example. Let $F = x^6 - 15x^4 + 8x^3 - 1$ and let's ask CoCoA to calculate the matrix Companion(F).

```
F :=  x^6-15x^4+8x^3-1;
Companion(F);
Mat([
    [0, 0, 0, 0, 0, 1],
    [1, 0, 0, 0, 0, 0],
    [0, 1, 0, 0, 0, 0],
    [0, 0, 1, 0, 0, -8],
    [0, 0, 0, 1, 0, 15],
    [0, 0, 0, 0, 1, 0]
])
-------------------------------
```

Now let's calculate CharMat(F).

```
A := CharMat(F);
Mat([
    [x, 0, 0, 0, 0, -1],
    [-1, x, 0, 0, 0, 0],
    [0, -1, x, 0, 0, 0],
    [0, 0, -1, x, 0, 8],
    [0, 0, 0, -1, x, -15],
    [0, 0, 0, 0, -1, x]
])
-------------------------------
```

We know that the characteristic polynomial of the matrix Companion(F) is the determinant of the matrix CharMat(F). Thus, let's verify that we indeed do have an equality between F and the characteristic polynomial of Companion(F).

```
F=Det(CharMat(F));
TRUE
-------------------------------
```

Now the reader can continue with any example of a polynomial in one variable. However, I probably should advise you not to take the degree of the polynomial to be too big... even computers can get tired...

10 Computer Commandments

1 Always use the binary system

10 Never use the symbol 2

Conclusion?

The examples of Chapter 8 have confirmed the fact that the real symmetric matrices have many incredible properties, e.g. they are similar to diagonal matrices something that one cannot say, a priori, for matrices which are not symmetric. However, what can we say about general matrices?

Usually we ended the chapters above with some questions, questions that would be answered in succeeding chapters. But, the book is finished and thus we will not be able to give answers to the question just posed. Mathematics, like life, teaches us to ask questions and the search for answers generates other questions and this steady stream of questions goes on for our whole lives. For now I would suggest that the reader be content in knowing that there is indeed an answer to the preceding question and that the tools used to answer it are a little more complicated than the ones we have seen so far. The words *canonical form, Jordan canonical form, rational canonical form* are some of the words that are used in an answer.

But, the more difficult the questions the more challenging it is to find the answers and mathematics is a true gymnasium in which to *go beyond*. Thus, I will propose one final question: do you think that the clouds are a limit to studying the heavens? Has the reading of this book given you the desire to go beyond?

e voi pesáte metà se piove

(palindrome dedicated to the clouds
from PALINDROMES OF (LO)RENZO
by Lorenzo)

References

[−] As was already mentioned in the introduction, if you open any web browser and type in *linear algebra* or *matrix*, or *linear systems* or... you will be overwhelmed by a miriad of web pages containing information about books, conventions, conferences, workshops and more, all connected to the topics covered in this book. Thus, I have decided to limit the bibliography very simply to the software package CoCoA and to a couple of my expository articles that might be useful and interesting to the reader as a stimulus to *go beyond*. Only one thing still bothers me: should this book "Linear Algbera for everyone" be included in the bibliography?

[Co] CoCoA: a system for doing Computations in Commutative Algebra, Available at http://cocoa.dima.unige.it

[R01] Robbiano, L.: Teoremi di geometria euclidea: proviamo a dimostrarli automaticamente. Lettera Matematica Pristem **39–40**, 52–58 (2001)

[R06] Robbiano, L.: Tre Amici e la Computer Algebra. Bollettino U.M.I.-sez.A, La Matematica nella Società e nella Cultura, Serie VIII, **Vol. IX–A**, 1–23 (2006)

when is your birthday?
October ninth
what year?
every year!

Index